STUDIES IN ANCIENT TECHNOLOGY

VOLUME I

STUDIES IN ANCIENT TECHNOLOGY

BY

R.J. FORBES

VOLUME I

WITH 40 FIGURES AND 10 TABLES

THIRD EDITION

E.J. BRILL
LEIDEN • NEW YORK • KÖLN
1993

This volume is a reprint of the second edition 1964.

The paper in this book meets the guidelines for permanence and durability of the Committee on Production Guidelines for Book Longevity of the Council on Library Resources.

Library of Congress Cataloging-in-Publication Data

Forbes, R. J. (Robert James), 1900-
Studies in ancient technology / by R. J. Forbes.—3rd ed.
p. cm.
Includes bibliographical references and indexes.
ISBN 9004006214 (v. 1: cloth).—ISBN 9004006222 (v. 2: cloth).
— ISBN 9004006230 (v. 3: cloth)
1. Technology—History. I. Title.
T16.F67 1993
609—dc20 92-39555
CIP

ISBN 90 04 00621 4

PRINTED IN MALTA BY INTERPRINT LIMITED

CONTENTS

PREFACE

When requested to prepare a second edition of my Bitumen and Petroleum in Antiquity, which had long been out of print, it seemed to the author that the time had come to publish certain essays on ancient technology which he had written. We can not hope to improve on H. Blümner's Technologie und Terminologie der Gewerbe un Künste bei Griechen und Römern, which excellent book is now over fifty years old and unfortunately deals with the classical period only. Since new evidence on classical technology and, above all, abundant evidence on preclassical technology has become available, but this evidence is spread over many periodicals and publications. A series of small booklets, each containing a number of essays on ancient technology, covering both the preclassical and classical period and demonstrating the continuity and the divergence, seemed justified. This book is the first of this series to describe certain phases of ancient technology, to point out the gaps in our knowledge and to try and use modern technology, philological and archaeological evidence to sketch ancient technology in such a way that it may be of some use to inform historians, archaeologists and philologists on the skill and the material world of the ancients. By co-operation many of the problems still unsolved could be unknotted and the author will be grateful for comments and additions to the material here presented. May these booklets promote discussion and research.

March 21st 1954.

R. J. Forbes

Haringvlietstraat 1
Amsterdam (Zuid)

PREFACE TO THE SECOND EDITION

The author has attempted to bring the literature up to date, but he has refrained from changing the original text unless fundamentally new data or facts had been discovered since 1954. He has gratefully made use of the helpful suggestions of several critics and he hopes that in the future such suggestions will again come forward.

September 1963

R. J. Forbes

Chapter I

BITUMEN AND PETROLEUM IN ANTIQUITY

Introduction

The extensive use of different members of the petroleum family is one of the most characteristic expressions of our modern civilisation. Since the rise of the petroleum industry, after the development of drilling methods from 1860 onward, the use of bitumen has been revived. During many centuries, throughout the Middle Ages and the Renaissance, the knowledge of the properties of this material, so extensively used in Antiquity, had slumbered or degenerated and bitumen was only remembered for its supposed magical or medicinal properties.

The impetus to our modern asphalt-industry must be ascribed to the Greek doctor Eyrinis d'Eyrinis, who in 1721 rediscovered the Neuchâtel rockasphalt (1).

This branch of the petroleum industry has been developed by the constant efforts of chemists and engineers during the last century to the effect that bitumen is now used for a wide range of purposes.

It is prominent in road-building, paper manufacture, roofing material and mouldings, while it is extensively used in impregnation, insulating or waterproofing layers, floors, tiles, paints and briquetting of coal-dust.

In a way, therefore, many uses of bitumen were rediscovered after a long slumber.

Everyone familiar with the results of modern archaeology will have found many references to the use of bitumen in Antiquity. It is certainly a common building material in the eastern half of the Fertile Cresent, and it is often mentioned in connection with mummification or lacquers in Egypt. At the same time its heighdays seem to be over with the coming of Hellenism. Amongst both the Romans and Greeks there is little in the way of the use of bitumens, but tars and pitches are then more commonly used. There is a natural explanation for these facts.

It is true that wood tar and wood-tar pitch are known ages before Hellenism, but the very frequent use of bitumen for all kinds of purposes involved the production of large quantities of this type of material.

Though there was of course a considerable production of charcoal for metallurgical and heating purposes (2), and thus a production of tar as a possible sideline, it is hardly to be believed that this quantity could have competed seriously with the large amount of bitumen, which could be produced comparatively easily from natural sources.

Then again the supply of wood for these purposes steadily waned as the forests of Syria and Mesopotamia were cut down in the course of the ages. Hence wood became too costly to be wasted and to be burned to charcoal and the production of wood tar progressively limited.

For the Roman Empire conditions were, however, reversed. The main sources of the supply of bitumen, the surface deposits of Mesopotamia were generally under the control of its enemies. The circumstances for the production of wood tar and pitch were, however, more favourable than in former ages, for not only the forests of Calabria and Thessaly, but all the wealth of timber of N.W. Europe was placed at the disposal of the disciples of the Brutti and the Macedonians.

Still this type of material gradually disappeared from general use and remained only in special branches as shipbuilding and of course in magic and medicine.

Thus the ancient knowledge of these materials was permitted to sink into oblivion.

The fantastic claims made for them by medieval scientists are eloquent for the deep slumber into which this knowledge had fallen till their rediscovery in Eyrinis' age. But then, when this knowledge and the significance of this group of products to technology and science were reborn, it would be rash to speak of a renascence, for there had been a considerable shifting of values.

Whereas in Antiquity the solid or semi-solid members of the bituminous family were valued and won above all, the significance of the lighter, more volatile types of petroleums and tars began to dawn upon the 17th and 18th centuries. Then the development of destillation apparatus, the kerosene lamp, deep well-drilling and the internal combustion engine contributed to the gradual predominance of these lighter products over their solid brothers.

Thus the present importance of the petroleum and the tar industries was born.

The lighter products were too volatile and dangerous to be of common use in Antiquity. The absence of proper distillation methods made it impossible to estimate them at their proper value. True, the lighter oils are constantly mentioned, but almost invariably in connection with their inflammability and, therefore, danger. This is well expressed in a passage from Pliny (3): "Some authors include naphta (as described in Book Two) under the category of bitumen, but its inflammability and close relationship to fire preclude it from any useful employment."

Very occasionally they are put to use for this very reason, until finally the Byzantines forge this knowledge into a new weapon of war, the well-known "*Greek fire*", but only after the development of early distillation technique had made this possible.

The object of the present essay is to bring together all the information on the use of bituminous materials in Antiquity. From the foregoing it will be clear that we will mostly discuss the solid or semi-solid members of this family; it is difficult to separate their story from that of the lighter products, which often explains many dark points of the former and which for that reason will be given too as far as possible.

Before summing up the literary and archaeological evidence it is necessary to discuss both nomenclature and classification of bituminous materials.

Nomenclature and classification

From the very first our subject confronts us with more or less serious obstacles.

One of the fundamental difficulties is the nomenclature of these bituminous materials.

The reasons for this embarrassing confusion are threefold:

1. The many records left us by classical writers or excavated at preclassical sites use the different terms for these substances promiscuously. The only distinction usually made or mostly at once evident from the context is that between the liquid and the solid or semi-solid bitumens. It is often extremely difficult to hit upon the modern equivalent for the terms "Asphaltos", "Bitumen" or "Naphta". The simple explanation for this fact is the lack of practical knowledge on these materials on the part of the classical writers, who left us the bulk of the existing data on this subject. Then again it was of course impossible to distinguish them analytically in Antiquity. Still several Sumerian terms seem quite clear and exact.

2. Though excavators' reports furnish us with a wealth of material, only very little scientific investigation has hitherto been applied to the products found, with the consequence that the enquirer has often to be satisfied with a vague description such as "asphalt" or "pitch" or "bituminous earth". Hence, precisely the point which is of very great interest from a technical point of view, namely, the kind of bitumen and the form or composition in which bitumens were used, is by no means always made clear. Here is an interesting problem, as yet practically untouched, open to further scientific research and sure to furnish interesting results for both archaeologists and technicians.

3. But even modern nomenclature is by no means a settled point. Though as far back as the beginning of the 16th century Libavius went into this problem and tried to define the different members of the petroleum-coal family (4, 5), confusion in modern scientific literature is as great as ever.

Still slow but persistent progress is made towards an international, well defined nomenclature by the constant efforts of Abraham (6) and Marcusson (7). The classification and nomenclature adopted by the writer is that of Abraham, though it differs in some details from the more international standards set by the *Dictionary of the Association Permanente Internationale des Congrès de la Route*.

Abraham's system is well planned and more logical, it suits our present purpose entirely and is more comprehensible to outsiders than the less exact terminology of the above-named *Dictionary*. This classification was used in composing a table, in which the ancient terms these materials were arranged, according to the substance, which they generally denoted (Table I).

The modern classification contains two groups of natural products (I & II) and two groups of artificial products (III & IV). The first group embraces practically all the natural derivates of crude petroleum and is therefore called "*Bitumens*" by Abraham. He defines this word as: "A generic term applied to native substances of variable colour, hardness, and volatility; composed principally of saturated hydrocarbons, substantially free from oxygenated bodies; sometimes associated with mineral matter, the non-mineral constituents being fusible and largely soluble in carbon disulphide (CS_2)."

The *Petroleums*, viz. all crude oils, are of a liquid consistency though sometimes by a lack of volatile components (gasoline, kerosene), they are semi-solid or very viscous. The crude oils with a high content of crystallizable paraffins (paraffin wax) are not very frequent in surface deposits or seepages of the Ancient World, they are less suitable for the production of asphaltic bitumen and can be practically ignored for our purpose. This limits the useful crude oils to the so-called "*asphaltic crude oils*", containing no crystallizable paraffine. However, the ancients lacked the distillation techniques to produce asphaltic bitumen from these crudes (8).

The Mineral Waxes are viscous to solid with a characteristic lustre and unctuous feel, the non-mineral matter consisting mostly of crystallizable paraffines. They do not occur very frequently, even at the present time only few rich deposits are known (Galicia) and they were probably unknown in Antiquity, though the term "ampelitis" is sometimes used for a similar substance as far as our meagre evidence goes. Far more important for ancient technology are the *Native Asphalts*. They are semi-solid to solid, thus showing considerable variation in hardness (often caused by the inclusion of finely dispersed water particles, viz. emulsified water), dark colour, comparatively non-volatile, containing little or no crystallizable paraffins; the non-mineral constituents being fusible and soluble in CS_2. Two different sub-groups may be recognized, viz. *Asphalts* and *Rockasphalts*, the only difference being, whether they contain less or more than 10% of associated mineral matter. The pure or fairly pure "Asphalts" are practically equal to the petroleum asphalts, derived from asphaltic crude oils by distilling off the more volatile fractions; sometimes they are so pure, that there is no analytical difference between asphalts and petroleum asphalts. Especially when mixed with mineral matter for technical purposes it is hardly feasible to distinguish them by analysis and we will therefore use the term *bitumen* for these two species, when we can not define our product more strictly as one of them.

The *Asphaltites* differ from the native asphalts by a higher fusing point and often they are either difficultly or only partly fusible. They are usually very hard and brittle, show a conchoidal fracture and generally contain only little associated mineral matter.

Thus their general characteristics preclude their frequent use in ancient technology.

Even nowadays they are only used for specific purposes such as bituminous lacquers, etc., where their high lustre can be put to use. They are pretty widely spread in small deposits, some of them occurring in the Ancient World. The *Pyrobitumens* embrace both the members of the coal family and those very rare substances called *Asphaltic Pyrobitumens*, representing the utmost stage of evaporation and oxydation of asphaltic crude oils. They have practically no value being infusible and insoluble in CS_2 and other solvents, the colour is dull and the brittle products give a black streak on porcelain. As they hardly occur anywhere else than in the American continent, they must have been unknown in Antiquity.

The *Non-asphaltic Pyrobitumens* represent the coal family. Of its members peat was only known in later classical times, lignite was probably known but hardly put to use as were the different types of coal in the Greek and Roman World.

The two remaining big groups of similar products are all artificial, they represent the distillates (Group III) or the residues (Group IV) obtained by distillation of crude oils or dry distillation of wood resins, gums or any member of the coal family.

The *Pyrogeneous Waxes* were certainly unknown in Antiquity. It was only during the last century that these products, commercially known as "paraffin wax", were prepared from petroleum distillates by suitable methods of cooling and filtration.

The *Tars* were certainly known, although a few of them only. For peat tar, coal tar and the like were unknown for reasons given above.

The dry distillation process for obtaining tar from coal is a recent development, which entered its commercial stage about 1790.

Wood tar was of course known. When producing the charcoal, necessary in several metallurgical processes, this substance is inevitably produced as a by-product. We shall have occasion to see that the Romans were well acquainted with its production, to the extent of selecting those trees for its production which gave the largest output and a long history must have preceded this technique. Therefore several words denoting wood tar may be found in ancient languages.

Similar more solid products are the *Pyrogeneous residues*. Our modern petroleum asphalts (asphaltic bitumen) form the subgroup *pyrogeneous asphalts*.

Their production in Antiquity from either asphaltic crude oil (by distillation or rather evaporation) or from rockasphalt (by melting and skimming off the molten asphalt) is very probable. We shall have occasion to discuss this point, but must remark here that it is impossible to distinguish them chemically from the *asphalts* already mentioned as occurring in a native state in Mesopotamia.

As both these petroleum asphalts and native asphalts are the same, chemically and physically, we will call them *bitumen* for our purpose

The pyrogeneous residues from tars are the well-known *pitches* so often unhappily confused with the bitumens. For the reasons explained above only wood-tar pitch was a common member which was produced in sufficient quantity to compete with bitumen.

This is not the case with those pitches, obtained by dry distillation of gums, resins, cedar oil, fat (fatty acids) which residues may have been known, but which were certainly too rare to be of any practical value.

Summing up our evidence we may expect to find in Antiquity crude oils, asphalt, rock-asphalt, petroleum-asphalt, wood tar and wood tar pitch as the most widely used members of the group of bituminous substances.

The coal family may have been put to use for heating purposes, but not in architecture, etc., where only bitumen, wood tar and pitch had the required properties to be employed in mixtures with mineral matter as mortar, plaster and the like. These artificial mixtures are generally called mastics, for which substances several terms have survived.

In the further columns of Table I we have endeavoured to enter the ancient terms for these products in their right place. Our leading principle was to take the most frequent meaning of such a word, as given by the study and interpretation of the texts in which it is used and to put the term in our Table next to its modern equivalent.

Ever since the hieroglyphics were first deciphered three terms have been taken by some investigators to stand for bitumen or bituminous materials, and translations in this sense and quotations in technical literature are often based on them. We shall now inquire whether there is any justification for this.

First of all there is the word *sift* (Table III). The Erman-Grapow dictionary (9) gives Salböl (ointment) for śft (the more modern spelling, as no certainty exists about the vowels in Egyptian words). Certainly, as far as we are able to judge by the texts, this was the most common meaning of the word before the Hellenistic period (300 B.C.). This word sift occurs even in the old pyramid texts, namely, in the list of ointments (10). It holds a prominent place in the religious literature as one of the seven (later eight to ten) "holy" oils. Reutter, who always took this material to be bitumen, assumed it, on the strength of prescriptions, to be present in several ointments prepared with these holy oils, and says that he identified it in such ointments, a statement which Lucas (11) rightly doubts. In this connexion it is certainly an oil, for

Table II

Metamorphosis and origin of oil, bitumen, asphalt and coals

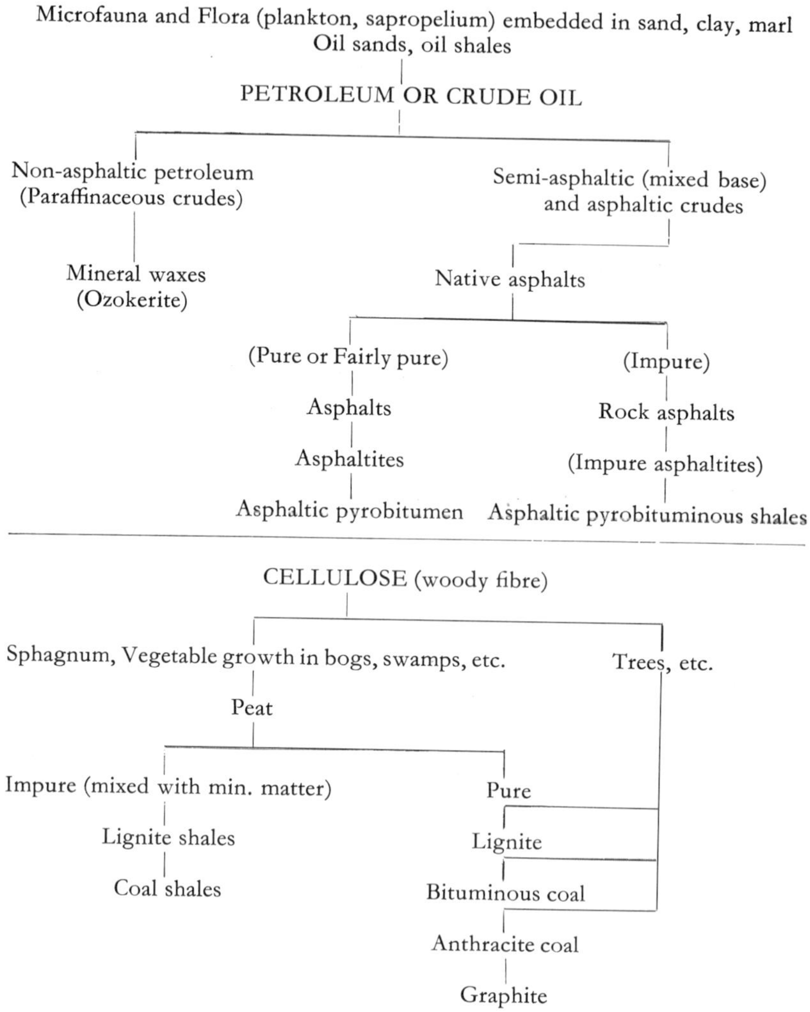

it is mentioned in one breath with other oils, such as cedar-oil, etc. A determinative denoting "oils" is often added to the word sift in the texts, as, for instance, in a very early one dating from the reign of Sahure, in which sift-oil is donated as a burial gift (12); also in a later text of the time of Pepi II (13) both, therefore, date from before 2000 B.C.

In several of what are known as the Later Egyptian texts (after

1580) Thotmes III mentions sift-oil as one of the tributes he brought back from his campaigns in Syria (14); and a pitcher full of sift-oil is mentioned in a somewhat later text of Rameses III (15).

Finally, when discussing a mummification ritual of the sacred Apis bulls (text between 250 and 100 B.C.), Spiegelberg comes to the conclusion that "here sift still clearly stands for an oil"; it has not yet acquired the meaning of sife derived from the Coptic (fig. 1, c) resin, pitch. It is not used for filling, but as an ointment (16).

In some texts there appears to be a certain connection with cedar-wood, and occasionally it has not been necessary to translate sift vaguely as Salboel, but more determinately as Cedar-oil pitch. Thus Sethe (17) says:

"Besides soda, cedar-pitch played an important part, though usually described as bitumen by the modernists. As long ago as in the Old Kingdom it was mentioned in connection with burial in a coffin made of cedar-wood from Lebanon. In the well-known Admonitions of an Egyptian Sage (about 2000 B.C.) the interruption of sea-borne traffic with Byblos is lamented because "neither the cedar for the interment of the mummies, nor its pitch (śft), comes to Egypt." (18)

Sift is given the meaning of turpentine(?) in a translation of the great Ebers medical papyrus (19).

All that can be gathered from the texts regarding the meaning of sift is, therefore, that it often came from Syria and was usually an oil, but that it can occasionally be translated as cedar-pitch. Cedar-pitch had been known for a very long time in Egypt, as Menghin's discoveries in prehistoric graves at Ma'adi (20) prove. There is nothing to show that it was bitumen. Certainly, in the later Coptic it means resin or pitch, as does also the Hebrew word zepheth, which is probably akin to sift. Whether this is synonymous with bitumen we do not yet know. Sift is also closely connected with the process of mummification.

Another old Egyptian word that has been thought to be related to bitumen is merhi (Table III, b). Erman-Grapow translated mrḥ.t (21) as Salboel, Oel, deriving it from a verb wrḥ meaning anoint, a derivation which Brugsch also gives (22).

About fifty years ago Maspéro devoted much study to this word (23). In papyrus No. 5158 at the Louvre and papyrus No. 3 in the Boulaq Museum, he found descriptions of the process of mummification in which different kinds of merhi are mentioned and came to the conclusion that the word stood for oils used "pour y faire dissoudre toutes sortes de matières qui la rendent épaisse et pâteuse" (to dissolve all kinds of substances which make it thick and pasty). In some instances he assumes that it means "poix (huile?) extrait des arbres croissant sur les montagnes du Liban" (pitch (oil?) extracted from the trees growing on the heights of Lebanon).

Loret subjected that study to close criticism (24), and came to the conclusion that the derivatives of merhi do not generally denote bitumen; the word merhi is always translated as oil in the general sense of ointment. He suggests, nevertheless, that this word has the same meaning as mennen, which we shall discuss, so that, according to his reasoning, it would denote bitumen after all.

Table III Nomenclature of Bituminous Materials in Antiquity

	Hieroglyphs	Coptic	Hebrew		Cuneiform
a	m.n.n.n				
b	m r ḥ.t	ⲁⲙⲣⲏϩⲉ amrehe	חֵמָר ḥêmâr	(Arab. alhummar)	
c	s̓ f t	ⲥⲓϥⲉ sife	זֶפֶת zépheth	(Aram. ziphthâ) (Arab. zift)	
d			כֹּפֶר kopher	(Arab. Kafr. Kir) (Syr. Kûphrâ)	ŠL 488 Ku-up-ru (rare) (The common sign for Kupru is ŠL 487.5)
e				(Syr. 'êtdo)	ŠL 487 ESIR ≡ iṭṭu = iddû
f					ŠL 484 (apsû) iddû ≡ ESIR
g					ŠL 483 (lagab)

Merhi (mrḥ.t) is a word of frequent occurrence in the Pyramid texts (25), where it signifies an oil used on a large scale, but it is not made clear whether it is of animal or of vegetable origin. Chassinat (26) quotes examples, ranging over the entire period of Egyptian history up to the Roman period, in which this word is used to denote an oil, certainly not a grease. Budge translates it as unguent, grease, fat of any kind, and Ebers as oil, animal fat, in any event liquid or unctuous substance.

At the end of his study Chassinat adds that in some cases the word may have some other signification, viz. "dans les références liturgiques suivant le cas de l'huile, de la graisse, de la poix ou du bitume" (malthe, poix minérale) (in liturgical references, oil, grease, pitch or bitumen (mineral pitch) as the case may be). There is, at all events, nothing to justify the translation of this word by bitumen unless it is clear from the context that this is the meaning, or in those cases where, in a bilingual text, it is made equivalent to terms definitely known to mean bitumen. For example, a late text calls it an equivalent of the Coptic amrehe, which means bitumen (27).

The third term, mennen (Table III, a) has more right to be translated as bitumen. Brugsch gave it (28) with reference to a text in the Pap. Rhind (vi, 1), in which mention is made of boiling (melting?) bitumen. Erman-Grapow (29) translates it as bitumen. Maspéro (30) points out to us that mennen was imported into Egypt from Phoenicia, Thabor or Punt (S. Arabia?) via Koptos, and thinks it stands for a kind of resin. Loret (31) had disputed this and thinks bitumen is meant. He admits that it comes from Syria or Phoenicia to Egypt and in some cases from Punt via Koptos, but bases his argument upon the known text of Dioscorides (32) in which he dwells on the use of bitumen for mummification. He also points out that the word mennen is often written with a suffix (so-called determinative) in the hieroglyphic texts, denoting granular or pulverized materials. He moreover cites the recipe for a scented ointment, which speaks of triturating and sifting mennen. It is also mentioned together with soda for the treatment of the swathes of mummies. Various quantities of mennen are mentioned in the great Harris Papyrus, e.g. 30 kilogrammes (xxxiii, B. 12), 32 baskets (lxiv, C. 12) and 30 kg, or 10 baskets (lxx, B. 14—15), in the last-mentioned case cited in one breath with quantities of certain resins. Yet Loret still doubts his own identification of mennen with bitumen.

It is also mentioned in a medical excerpt of the Ebers Papyrus (54. 19) which Ebbell translated (33). It contains a prescription to triturate orpiment together with bitumen (mennen) and to place this mixture on heated bricks. It claims that inhalation of the vapours will relieve a cough!

We are inclined to infer from all these data that mennen stands for the glancepitch known especially in Syria. This material greatly resembles the resins so well known to the Egyptians, as far as appearance and melting point are concerned, and they therefore perhaps mistook it for a resin. This is not unlikely, seeing that mennen was used for all the treatments involved in mummification for which resins were also employed, and so perhaps Maspéro and Loret are both right in the end. One thing is certain, namely, that mennen is explicitly made equivalent to the Coptic amrehe in later texts and that it came from those parts where deposits of glancepitch are known to be, which deposits seem to have been mined very early (34).

One of the best-known terms for petroleum is the Akkadian word napṭu (Hebr. naṭpik; Greek, *νάφθα*; Arabic, alnaft). The source of this word has long been a matter of doubt. Many investigators have taken it for an Iranian word. This is Darmesteter's opinion (35), for instance, who derives naphtha from an old Iranian word nab (to wet); which he then assumes to have passed into the Persian language of the Avesta's as napta; the latter word is then supposed to have given rise in Mesopotamia to the Akkadian word napṭu.

At the Second World Petroleum Congress (Paris, 1937) Herzfeld not only defended this theory, but even quoted from the myths of old Iran (36) to support it. Although the word naphtha is not mentioned anywhere in these myths, a hymn to the goddess of the Earth speaks of an eternal fire, which needs no feeding and is to be found on the southern coast of the Caspian Sea. Another hymn mentions a fire—a symbol for Iran—burning in the ocean. This is a legend which also has come down to us from Indian lore. Further, the Iranian myths have a water-god, Apam-Napat, who has much in common with the god Neptune. Attempts have been made to relate the names of these two gods to the word naphtha, or at any rate to the original stem nab (Meillet). We cannot go any further into the philological aspect of this problem (on which expert opinion is still divided), but may point out that, as the word naphtha has not been expressly mentioned in these myths, the connexion should not be assumed to exist as a matter of course. These mythical fires may be nothing more or less than different forms of the sacred fire which plays such an important part in Iranian religious cult. Supernatural qualities or magic powers were ascribed to these fires, but there is nothing to show that they were mentally connected with burning natural gases or petroleum. Fires invested with magic power occur in the legends of practically every people in whose religious cult fire plays a part. However, we possess other proofs of the existence of the word napṭu in Mesopotamia, before there can be any question of an Iranian text. Its derivation from napâṭu, nabâṭu (to flare up, to blaze) is in perfect agreement with Akkadian habits of word formation and proves that here the word napṭu was not formed for a special application, but described the natural phenomenon, like other terms for bitumens in Mesopotamia.

We moreover possess a number of old texts containing this word. The most important is an old Babylonian text (37) which, according to Hunger (38), dates from about 2000 B.C. and in which napṭu is used in connection with oracles. We would further mention four texts, of which, admittedly, we only have a later version, but which certainly draw on older originals. In one of them a certain person pays $\frac{1}{6}$ shekel of silver for naphtha (39). Two omen-texts speak of a flood of naphtha (40) or a naphtha fire in a certain part of the town (41) as ill omens. In a fourth text it says: "If in a certain place of the land naphtha oozes cut, that country will walk in widowhood" (42).

These few lines are probably sufficient to rule out any but an Akkadian origin for this word and to show that it spread from Mesopotamia both to Iran and westward. That is why we cannot agree with Séguin, who at one time thought he had to do with a Hebrew original (43). In Macc. II, i, 36, naphtha is mentioned in direct connection with the "Persians". A curious point about all these texts is that naphtha still stands for crude oil. It was not until after the introduction of distillation that the word naphtha gradually came to mean gasoline.

The classics use naphtha for "crude oil" generally, maltha being sometimes used for a very thick and viscous asphaltic crude oil, though its true meaning seems to be the pure native asphalt which we will discuss later on. Latin texts often call it bitumen liquidum, though they generally use the terms of the Greeks freely.

The word petroleum was not coined until the Middle Ages and then used for crude oil. Later, when the fraction boiling at somewhat higher temperatures than gasoline or petrol, was also distilled off and used for lighting purposes the term "kerosene" came into use. But these and other fancy names for petroleum products have no etymological connection with the earliest terms. The transition in the meaning of naphtha is quite clear in the writings of al-Kazwînî who speaks of alkir as a synonym of malta and of black al-nafṭ (naphtha, crude-oil) and white al-nafṭ obtained from it by distillation (and hence denoting petrol or similar fractions).

The other terms for the asphaltic crude oil are directly related to the bitumen which can be prepared from it in such large quantities, a relation, therefore, of which the Ancients were already fully aware.

The term šaman-iddî, meaning literally oil of bitumen, was used for crude oils. Šaman-šadî (mountain oil) is a term that also occurs. But even older names are known to us. The Sumerian term for the former is IÀ-GIS-ESIR (44), the latter is called IÀ-KUR-RA (45). Mention is also made of ESIR-NE (literally "fire bitumen") (46) which probably refers to burning natural gases or, perhaps, to crude oil.

Of much more importance to the Sumerians, the oldest inhabitants of Mesopotamia, were the solid and semi-solid bitumens, and this is clearly shown by their nomenclature, which covers a variety of names for grades of bitumens. At the same time it is evident by this nomenclature that bitumen was mainly obtained from seepages. True, rock asphalts and asphaltites were known and sometimes used, but in production and economic importance they were far behind the other products. But the Sumerians seem to connect definitely "bitumen" and (subterranean) water. One has only to consider Table III, c, which is descended from the oldest cuneiform sign for ESIR (bitumen), to see that the latter conclusion is justified. Of the two signs used for ESIR, Table III, f, occurs invariably in the oldest texts, but is represented in the later texts by Table III, e, which more particularly means "bitumen". However, both characters are composite signs or ideograms and go back to the basic pictograph LAGAB (Table III, g) according to Deimel (47) which originally meant "cask", and was later on also used for "surround, shut off, etc.". By combining it with other signs the two ideograms for ESIR were formed. Thus the oldest sign is taken to be a combination of LAGAB + ḪAL, a combination which we do know from the oldest pictographs (48) in the form given in the figure, but which, in cuneiform, had degenerated to ŠL, 484. This symbol not only has the general meaning of "ocean, fresh water, abyss", but is also used for "river, canal, source", and similar concepts. This has something to do with the Sumerian conception of our world. They conceived of it as floating on a large freshwater lake, the "Apsû", from which rivers and wells received their water and from which also rose all kinds of demons to ensnare mankind.

Hence a symbol was used for bitumen indicating that this substance, like river-springs and well-water, oozes from the depths of the Earth. Very often this symbol is found combined with the symbol for water, accentuating, as it were, the idea of bitumen being a substance issuing from the freshwater abyss, or as Deimel expresses it: "Wie das Süsswasser quillt in Mesopotamien das Bitumen aus dem Apsû hervor" (ŠL, 457). (In Mesopotamia bitumen wells up, like fresh water, out of the Apsû).

We have to wait until we get to the late Sumerian texts to find a

Fig. 1.
Bitumen spring near Hit.

separate sign (Table III, e), a compound of LAGAB and NUMEN, which we do not know from old Sumerian texts (ŠL, 487), but which is generally used for ESIR. The later Babylonians and Assyrians, who spoke a Semitic language (Akkadian), but used the old Sumerian signs, to which, consequently, they imparted a phonetic value, compiled many vocabularies, a kind of Sumerian-Akkadian dictionaries. From these dictionaries it is quite clear that this late-Sumerian sign is used for the Akkadian words iddû and kupru, both meaning bitumen. We shall revert to these terms later on (49).

In the texts bitumen is written in two ways, viz., ŠL, 487 preceded by the symbol for water; this combination is represented by ESIR. The symbol ŠL 487 also occurs alone, and is then represented by ESÌR (or $esir_2$). We shall adopt this (Continental) spelling in our subsequent pages.

This relation between bitumen and the freshwater abyss also appears from other examples, e.g. in magic texts, in which bitumen is the seat or the symbol of powerful evil spirits, spirits rising from the Apsû to do harm to man.

In several vocabularies bitumen is related to other products deposited by the water on the banks of the rivers. The Sumerians were acute observers of natural phenomena around them, and the peculiar structure of their language enabled them to express the qualities (especially visual ones) of rocks, minerals, plants and animals, etc. in their nomenclature. Thus, they used the same rootword for a group of objects, e.g. rocks which they thought were related, and differentiated each member of such a group by a suffix expressing its characteristic properties. Thus we see the products from the rivers (Sumerian, ÍD) grouped as follows (50):

(Sumerian)	(Akkadian)	
KI-A-(AN)-ÍD	kibritu	black crude sulphur
ÚH-(AN)-ÍD		yellow sulphur
BA-BA-ZA-(AN)-ÍD	pappasi	gypsum from the Euphrates
A-GAR-GAR-(AN)-ÍD (ESIR)	(iddû, kupru?)	bitumen, pitch.

In a similar way various bitumens are distinguished by suffixes to the root ESIR, to which we shall revert. The A-GAR-GAR-(AN)-ÍD occurs usually in medical texts and, according to Campbell Thompson, seems to be the crude wet form of bitumen obtained from the seepages.

As to the word ESIR, HAUPT (51) has pointed out that it stands by a common vocalization for a stem ASIR, lit. illuminating water(?), Besides, ESIR or iddû is mentioned positively with special reference to the river god ÍD (52).

All this is given tentatively as additional proof that seepages in the vicinity of Hit played a leading part in the production of bitumen in ancient Mesopotamia. The present town of Hit is often mentioned in Sumerian literature as Duddul (Duldul, Dudul) or Idu (53); later on the Greeks refer to Is. The word Duddul is assumed to be derived from DUL (well), and Idu (I-TU) seems to be connected with ÍD (river).

This town being the principal production centre of bitumen, its later Akkadian name iddû was given to bitumen (54) so that the present name Hit may be considered as being derived from the Sumerian name. This iddû, also written iṭṭû or ittû (55), therefore first indicated the product from the seepages; later on it was used in a more general sense.

If we want to define more narrowly the different kinds of ESIR occurring in Sumerian literature, we must not, like Forrer (56), start exclusively from the question as to whether these different kinds are measured or sold by weight or by volume, as this reasoning may lead to wrong conclusions. It is far better to ascertain in the first place for what purposes the kinds of ESIR mentioned in the texts were used, taking into account at the same time whether they were sold by weight or by volume, and, of course, not neglecting the price mentioned for the various grades. The so-called Period of the Third Dynasty of Ur (ca. 2100—2000 B.C.), Ur III, provides much information about these prices, enabling us to arrive at an average price. We have calculated these prices in shekels (Sum. GIN) (ca. 8.5 g of silver) per ton; in the absence of exact data by which to fix an index figure for these early times, these prices may be considered only as relative values, but this is all we need for our purpose.

It is permissible to draw conclusions from the Sumerian nomenclature directly, provided they be confirmed by the texts concerning applications, as the translation for the various names is by no means firmly established in all cases. A point which is naturally confusing to anyone not acquainted with the Sumerian language is the fact that various scholars pick out different values for one cuneiform sign; thus their translations show different terms which are in reality the same. We will give the terms as found in the papers cited in each case.

Let us first take the group of natural asphalts. Those obtained from seepages were, of course, of the utmost importance.

ESIR-LAḪ or ESIR-UD (UD, PAR (see below), and LAḪ are indicated by the same ideogram) seems to be the natural asphalt from the seepages, which in purity and freeness from ash approaches present-day asphaltic bitumen. Only small amounts of occluded water have to be removed from it. We are inclined to infer from the texts at our disposal that ESIR-LAḪ is this purified bitumen, which, therefore, in point of properties, corresponds to the asphaltic bitumen obtained from crude oil by inspissation, or from rock-asphalt by melting under certain conditions.

In the Ur III period we find many lists of large quantities of ESIR-LAḪ expressed weight or volume, in quantities varying between 4950 and 20 kg, and between 8500 and 1089 litres (average ca. 2900 kg); the price for quantities by weight and volume is the same, i.e. roughly, 3—5 shekels per ton. One text even speaks of a tribute of 280 tons of ESIR-LAḪ to be paid to the king of Ur by the town of Girsu (57). This bitumen is frequently mentioned in connection with brick-work (inter alia "for the house of the Grand Vizier") (58) and with shipbuilding (59). Another text of the same period tells us that five ships laden with ESÍR-LAḪ have been despatched to Ur.

The term ESÍR-PAR is also mentioned (60), but as PAR is one of the values of the ideogram UD, this is identical with the above grades. Our final conclusion is, therefore, that translations of texts mention the following different readings of the same cuneiform signs for the purified bitumen from the pools of Hit: 1) ESIR-LAḪ, 2) ESÍR-UD, 3) ESÍR-PAR.

For the crude impure product of the pools we already mentioned the terms A-GAR-GAR-(AN)-ÍD or A-GAR-GAR-(dingir)-ÍD, which seems to be used in medical texts in particular. Besides this term, however, ESIR-UD-DU-A, ESIR-È-A (61) is also used. The addition È-A (ŠL, 579, 458) means, according to THUREAU-DANGIN, qui sort, qui jaillit (which emerges from, springs from), so that we are here clearly concerned with the crude bitumen which oozes up near Hit. It is identified by several scholars with kupru (62). This kind of bitumen is mentioned in medical texts (63) and seems to have been used for the seams of ships. Hence, the crude natural asphalt is called 1) ESIR-È(UD-DU)-A, 2) A-GAR-GAR-(AN)-(dingir)-ÍD.

Several terms are also known for rock-asphalts, the principal being ESÍR-ḪURSAG (64).

This word ḪURSAG means "mountains", perhaps more correctly mountain-range, in contradistinction to KUR, mountain. Obviously, therefore, rock-asphalt is meant in this case. Gudea, the priestly ruler of Lagaš (ca. 2400 B.C.) had it brought from the mountains of Magda. A quantity of 109 tons(?) is inscribed on the plaque bearing his image. It remains to be seen whether this is the exact quantity, or whether it does not rather stand for the whole consignment of bitumen, gypsum, etc., as the other texts of the slightly later Ur III period mention only smaller quantities, such as 7200 and 840 kilogrammes.

The other terms for rock-asphalt and like products were ESIR-A-BA-AL-ḪURSAG and ESÍR-A-BA-AL.

It is not clear as to what the termination A-BA-AL (65) really signifies. Possibly it means that the crude rock-asphalt boiled with water yields bitumen, but so far no account of this technique has been found in old writings. In BARTON's opinion it means(?) drawn or dipped from a well. Maybe, however, bitumen was obtained from this rock-asphalt by the very ancient process of destillatio per descensum (dripping) (66), as we have occasion to prove.

Like the rock-asphalt ESÍR-ḪURSAG, this rock-asphalt was sold by weight. Amounts like 606, 1515, 91, 45 kilogrammes and 55.3 tons(?) are mentioned on tablets of the Ur III period, so on the whole for smaller quantities than of the other group of bitumens, although the price is roughly the same, namely, 4.1 shekels per ton (67). This bitumen was used for the building of houses and terraces.

It was by no means only Gudea who obtained this bitumen from the mountains, for Sargon also imported rock-asphalt from Kimaš (Elam, the mountainous district along the present frontier between Iraq and Iran), and Ibi Sin had it brought from the region of Magda.

There was another product, akin to rock-asphalt, known by the name of IGI-ESÍR. This also came from the mountains (68) and was sold by weight (69). The texts of the Ur III period record only small amounts of 75 to 5 kilogrammes, an average of 40 kg, but the price is appreciably higher than that of the grades hitherto discussed, viz. 10.4 shekels per ton or more (70). THUREAU-DANGIN gives the word ESÍR-IGI-ENGUR in Gudea's big inscription (71). He assigns to this word the same meaning as KUNIN (ŠL 487,5) which, in turn, is identical with kupru and, as pars pro to, is used for a ship caulked with bitumen or bituminous mastic.

In view of the small quantities and high price, it may be assumed that the product in question is a bitumen (épuré) prepared by melting down rock-asphalt. The relatively slender demand for it may be explained firstly by the importation of large amounts of ESIR-LAḪ and, secondly, by the low yield and high prices of fuel.

To sum up, then, the following names have been found in translated texts of rock-asphalt: 1) ESÍR ḪÛRSAG, 2) ESÍR-A-BA-AL, and IGI-ESÍR for refined rock-asphalt (épuré?).

The names of other kinds of bitumen clearly express the use to which they were put. The product mentioned most often is ESÍR-É-A, which we will identify with bituminous mastic. Forrer questions this identification; he points out that, as this product is always referred to by volume, it must have been liquid. Moreover, he does not seem to distinguish the signs ESÍR-É-A from the ESIR-È-A already mentioned. (72). É is the sign for house (ŠL 487, 3) ESÍR-É-A is always mentioned in connection with buildings and brick-work ("for Urnammu's libation table"; "for Gimil-Sin's house"; "for the new palace", etc.) ("for a gypsum container"), with caulking ships, water proofing baskets, wicker, mats and the like (73). No fewer than 77 texts of the Ur III period mention this product; the amounts range from a few kilogrammes to 3500 kg, an average of 452 kg. These amounts are usually given by volume (74). This does not prove that the material was fluid, least of all if we remember that the liquid mastic mixture was poured into "loaves" or baskets. We have had occasion to examine various mastics composed of ESIR-LAḪ, fillers and fibrous material. When these were poured into the proper baskets or moulds, cakes were formed of a certain weight and volume. A sample originating from ancient Ur still clearly bore the marks of the basket in which it had been packed. The price of this material was considerably higher—at an average of 18 shekels per ton (21 shekels at a somewhat earlier date)—than that of the other kinds, a further proof that it is not the crude bitumen taken from the pools. Are we to assume that, fuel being very costly, it was the practice to mix the mastic on the site to avoid the second heating required for ready-mixed mastic? No wonder FISH says that this mastic was dearer than dates or barley! This high price was also no doubt responsible for the fact that very much smaller quantities of bitumen in this form were used than of the raw material for mastic. The translated texts mention other kinds of mastic as well, one of which was called ESÍR-GUL-GUL (or SÚN-SUN), the determinative of which term refers to machinery for irrigation; the word is found in a text at the Louvre (75) and others (76). Apparently this product was also used for caulking ships, ESÍR-APIN (or ENGUR) is mentioned for a similar use. This APIN (77) stands for sowing-plough, dredger and other apparatus used for irrigational purposes. This grade of ESÍR is also mentioned in connection with shipbuilding. The texts mention small amounts of 300 to 5 kilogrammes, but no prices (78).

The analyses of old samples, to be discussed later on, also go to prove that various kinds of mastic were made. They showed that the mortars for brick-work contained on an average 35 per cent of bitumen, and the mastic asphalt or asphalt mastic of floors and thresholds on an average 25 percent, a difference which, considering the number of samples examined, can scarcely be regarded as fortuitous.

An object is referred to in some texts under the name of GIRRA, and is said to be made of bitumen (79). FORRER relates this word—to our erroneously—with the present Arabic word qir. The texts should be scrutinized again by experts. Perhaps, as the price seems to indicate ESIR-GIR (UD) is actually meant.

A special kind of bitumen also occurs in Assyrian statute laws (80), i.e. GIR_4 (81), which, apparently, was imported from the north. DEIMEL mentions that the word was used in connection with furnaces for the melting of asphalt, pitch etc. By Assyrian law some delinquents subjected to corporal punishment had, in addition, a jar of molten GIR$_4$ poured over their heads. A similar punishment was meted out in the Hana district (82).

The following expressions are also interesting with regard to the use of bitumen:

The term ESIR-ŠUB-BA occurs in some texts of the Ur III period in connexion with house-doors and various objects, not yet clearly defined, made of reeds and rushes. This indubitably means "coating with bitumen". Delitzsch translates this as "malaxer du goudron", which certainly does not fit in these texts. Deimel and Meissner give approximately the same interpretation (83); Burrows (84) shares our view.

It is interesting to note that three Ur III texts speak of a ḪUM ESIR, a mortar for bitumen; but, unfortunately, they do not define the kind of bitumen. It is very probably rock-asphalt, pulverized for further treatment.

The number of grades of bitumen mentioned in the later Akkadian texts is limited to two. We used, certainly, to think amâru (85) was recognizable as a word for bitumen that might be the prototype of the Hebraic hêmâr, but closer study of the text proved that this was not so.

The word iddû is written with the symbol ESIR. As we have already contended, the word is probably connected with the name of the city, Hit, and first, therefore, stood for the soft, sometimes moist, product of the pools (86), but was afterwards used in a more general sense.

The second term, kupru, was usually represented by the symbol for ESIR. It is related to a root kpr, which means to coat, and is used as a rule for harder bitumen or mastic. There are two other signs (Table III, f) which are read as kupru; the first is mentioned by Deimel (87) and Delitzsch, the second by Barenton (88) and Delitzsch, and both are made equivalent to ESIR. So although a difference undoubtedly exists between these two kinds, as is apparent from several texts in which they occur side by side (89), this distinction is not always meticulously observed, especially in the later texts. The nomenclature suffered from the same carelessness in antiquity as in modern publications. As a rule iddû is rendered by asphalt, bitumen, etc., in translations, and kupru by poix, Erdpech, pitch, by which at least the difference in hardness is to some extent indicated.

There is certainly no Akkadian original of the Greek asphaltos, such as the hypothetical aspaltu which some technical handbooks mention. Lidell-Scott is right in deriving asphaltos from the verb "sphallo", to split. The Greek language possessed another word for these native asphalts, viz. pissasphaltos which is used especially for rock asphalts, though later writers often use it for fossil resins ampelitis (sometimes ozokerite or such mineral wax) and other vaguely similar substances which the Greeks with their much smaller knowledge and application of these materials failed to keep apart. Maltha is the purer asphalt and this term is taken over by the Romans, who, however, preferred the word bitumen.

"Lapis bituminis" occurs only in some post-classical alchemical writings, but seems to have used for our "rock-asphalt", as far as one can judge from the often obscure context. The Sanskrit words are given only tentatively. They are somewhat rare as are the native asphalts in India though a few pools of a thick black asphaltic crude oil exist in Northern India. However, these products seem to have been rather uncommon, we can deduce this too from the fact, that substances like pitch, asphalt(?), gum or resin were all expressed by the same word "jatu".

The vagueness of ancient nomenclature is very prominent in the words for the different members of the coal family, this is of course partly due to the lack of discrimination and the rare use of these materials. For heating or metallurgical and other purposes charcoal was the common fuel and the words for this material are very often used for coal too. Thus, for instance, "anthrax" is used in the sense of carbuncle by Aristotle, for charcoal by Thucydides (90) and for real coal by Theophrastus (91).

In the Middle Ages the confusion is greatly increased because of the indiscriminate use of "ampelitis" for all black fossil materials, whereas the ancients had used it only to denote rare asphaltite-like material (92) or like gagates for some kind of lignite (93). True coal was of course hardly used before the Middle Ages except perhaps for domestic use in Roman Britain (94). The general fuel was wood, camel dung or charcoal.

Terms, denoting wood tar and wood-tar pitch are much more frequent. Prof. Campbell Thompson mentions several in his publications (95) and other Akkadian terms are common on contract tablets (96). The Hebrew "kopher" and "zephet" occur very often in the Bible (97) and the classical words are very common in ancient literature. It must be borne in mind that incomplete as this table may be, such compilations and classifications are urgently needed.

Where bitumens were found

Our second point will be to find out where bitumens were found in Antiquity.

Complicated as this question may seem, several factors conspire very happily to lighten our task in answering it. The first of these factors is the development of *ancient mining methods*. It is well known that these methods were still primitive and were practically limited to placer-mining, open-cut mining or pit-mining. The exploitation of oil pools or asphalt-lakes may be called a form of placer-mining, they were of course easy enough. The open-cut way of digging out layers of bituminous rocks may have been used, but pit-mining, generally restricted in Antiquity to depths of less than 300′ must have been used very rarely. Where surface indications did not exist it was impossible for the ancients to have guessed the situation of an oil formation in the ground beneath.

Thus the second factor is this: if *ancient geological knowledge* was slight the exploitation of bitumens must have been limited to obvious deposits, i.e. to surface deposits. Indeed the only classical writer with a wider view on minerals giving exacter descriptions was Theophrastus (98).

Though Assyrian texts contain some remarkable practical observations on this subject, there is nothing worth noting on bituminous materials. We feel that though the general characteristics of these products were known, no real geological knowledge about their origin and deposits existed.

Again our way of tapping crude oil deposits by *deep drilling methods* could not be used, because technique was not yet sufficient advanced in ancient times and gas pressure usually existing in oil deposits made it impossible to follow pit-mining methods.

True, the Chinese seem to have used drilling since about 200 B.C. and have made wells to a depth of 3500′ with equipment of a surprisingly primitive nature (bamboo tubes and bronze bits). They have thus tapped oil and gas layers and used them for different purposes (heating and lighting), but this knowledge never reached the Old World in the period we are discussing (99).

We are thus led to the conclusion that it must have been impossible to exploit the deeper and richer bituminous deposits in Antiquity and that development was confined to the so-called surface deposits.

What is the nature of these "*surface deposits*"?

Crude oil or bitumen occurs in deposits of sandstone, limestone or clay (shale), where it is formed from the original plankton, which was deposited together with the sandstone shale or limestone. The oil-bearing stratum thus holds the crude oil in its pores together with the gas and water also formed by the chemical decomposition of the original plankton (Table II).

The gas formed during this prolonged chemical reaction often causes a high pressure in the oil-stratum and would have driven out the oil and water, had this not fortunately been prevented by impervious layers (clays, shales and the like) which usually shut off the oil strata from higher and lower layers. Still these sealing "cap-rocks" are often damaged by the earth's movements and the ensuing fissures can be a means of escape for oil, gas and water to the earth's surface. There pools are formed, so-called "*seepages*", which may still produce large quantities of very inflammable gases, which bring oil or bitumen up to the pool from the original oil-deposit. Seepages of this kind are very common in Mesopotamia. In Northern Persia and around the modern Baku we often find gas escaping alone through fissures in the earth. These *gas-wells* have been known as burning pillars of fire since untold ages. Even up to the present day we find Parsee temples near Baku.

However, several oil-strata do not remain underground but have one wing extruding from the earth's surface. When such an outcrop occurs it is natural that the gases escape and very often the lighter fractions of the enclosed oil evaporate slowly. The result is that the heavier residue remains in the rock. These outcrops of *bituminous rocks* or rock asphalts occur all over the world, their bitumen content varies from 4 to 20%, and we shall see that it is quite possible that bitumen was obtained from them in Antiquity.

Veins of asphaltites and asphaltic pyrobitumens are somewhat uncommon but still they belong to the surface deposits which may have been exploited.

Now that we know the nature of these surface deposits we must look for their occurrence in the ancient world and compare our modern knowledge of these outcrops with the references to them in ancient literature.

Africa is a continent relatively poor in bituminous materials and this seems to have been the case in Antiquity too.

One of the problems to be raised is that old controversal question as to whether the ancient Egyptians were acquainted with petroleum and bitumen and whether these materials were generally used in Egypt. One of the things which Séguin (100) has tried to show is that the frequent mention of mysterious, earth-fed fire which water will not quench reflects memories of, or acquaintance with, petroleum and natural gases in Egypt. An assumption of this kind cannot be accepted off-hand. Fire plays so eminent a part in the religion and magic art of nearly all ancient peoples that there is no justification for connecting a priori the many references in their texts and legends to mysterious fires (often of magic, i.e. chastening, power) with the burning natural gases or like phenomena known to petroleum geology. Flinders Petrie's theory (originally Fessenden's), quoted by Séguin, to the effect that the Egyptians originally came from the Caucasus and had memories of the natural gases and petroleum still burning there, was not only rejected by the archaeologists gathered together in Paris (1937), but was declared years ago by Petrie himself to be untenable. The vague knowledge that at one time bitumen was in some way connected with the preservation of mummies in Egypt is so exaggerated and stressed in most works on bitumen that this is often the only use of bitumen in antiquity the layman remembers.

Every investigator must realize that the old Egyptian texts give him little to go by. Although we have access to many medical texts, including several describing the process of mummification, as a rule the enumeration of the ingredients used is not clear enough to enable us to decide what substances were meant. So-called technical texts are not infrequently merely lists or prescriptions which make identification impossible, and so we have only gradually penetrated into the secrets of ancient Egyptian pharmacology.

So even though a few words in the texts may sometimes signify bitumen, and one almost certainly does, we have reason to believe, on other grounds as well, that bitumen was not an important factor in Egyptian life.

The Egyptians seldom availed themselves of the opportunity of collecting crude oil from the seepages on the coast of the Red Sea and producing bitumen from it. Recent investigations at the Djebel Zeit have shown that the seepages were not worked until Roman times when it was called Mons Petrolius and then only on a modest scale. Near the sea coast there are pits 3 metres deep containing many remains of the earthenware utensils with which the oil was skimmed off the pools (101). Slight amounts seem to have been used in pharmaceutical recipes. Such medical literature as the Papyrus Ebers (1550 B.C.) sometimes mentions pr ḥr ḫ3śt.f: "what oozes forth from the desert" or mrḥt ḫ3śt, "oil of the desert". From its applications in unguents and remedies for eye-diseases of the eyes it is clear that crude petroleum is meant, which is also used by the generation of Dioscorides and later pharmacists. It is interesting to note that these Egyptian terms are formed in the same manner as our present "petroleum"!

Modern forms of analysis have also failed to produce much proof of the use of bitumen in Egypt. There were also much richer seepages on the Sinai Peninsula near Abu Durba and Gebel Tanka, which do not appear to have been put to account either.

Not long ago, in the heart of Egypt, at Helwan near Cairo, strata of a fairly rich bituminous sandstone were discovered which appear to have been unknown before. Dioscorides mentions a special kind of material which he calls Memphites lithos (102). From the meagre description we are led to the conclusion that he must mean some kind of rock asphalt or perhaps an asphaltite. On the other hand the term "stone of Memphis" may have no direct connection with the former capital of Egypt. Although Joachim (103) translates the term "inr spdw" as "stone of Memphis", but this is incorrect, the literal translation "sharp stone" is the only one that fits the passages in the papyrus Ebers. There is therefore no reason to connect it with the Memphites lithos of Dioscorides. The total lack of references to sources of bitumen in ancient Egypt, when coupled with modern evidence, makes it pretty sure that bitumen was imported from Palestine or Syria. The only text mentioning bitumen from Hit (which will be referred to later on) in Karnak seems to record an exceptional case. In fact both Strabo (104) and Diodor (105) mention export of the Dead Sea bitumen to Egypt for the purpose of embalming. Though this use may not be as common as it would appear from these texts, the evidence must not be ignored.

Vitruvius (106) says: "There is also a lake of Ethiopia which anoints men who swim in it", but no modern oil-seepage in this country can be indicated as the source to which Vitruvius is referring. The same holds true for another passage in his book (106): "There is also a spring at Carthage on which floats an oil with the perfume of cedar shavings and with this oil sheep are usually dressed". The same seepage seems to have been known to Aristotle (107).

However, this seepage in North Africa can have been of local importance only, for we know to-day of several other seepages in this region but even modern drilling methods have failed to strike profitable oil-deposits except in the desert regions.

Palestine was a region much richer in surface deposits than Africa and we have a lot of reference on this country in ancient texts.

Seeing that from Ptolemaic times a certain amount of bitumen was exported from Palestine and Syria to Egypt, it is very strange that although the numerous deposits of bitumen in various forms were known even in antiquity, the bitumen does not seem to have been used in the countries of their origin themselves.

In excavating the remains of Jericho Garstang (108) found a 4′ thick wall enclosing an area of about 4 to 5 acres. It was built by cementing large bricks with bituminous earth and could be dated back to the Early Bronze Age (2500—2100 B.C.).

Albright (109) points out that a certain amount of bitumen has been found in various places. Thus Duncan found a "considerable deposit of lumps of asphalt" in Ophel, in Canaanite Jerusalem, dating from the Third Millennium, and in Tell Beit Mirsim nearly every stratum contains lumps of crude bitumen (from the Dead Sea?). These deposits most commonly date from the Second Millennium, but some from the Second Iron Age, i.e. 600—100 B.C. also. Albright thinks that the bitumen many possible have been used for making furniture, but that any traces of it there may have been disappeared with the remains of the wood. In view of the properties of bitumen as preserved in very old samples, we think that this is unlikely. It would undoubtedly serve a good purpose if the bitumen of these deposits were subjected to analysis.

Modern research has proved the existence of many surface deposits of bituminous limestone as, for instance near Hammath, Tiberias and on the east shore of the Dead Sea. It may be that they are represented in a passage of Vitruvius (110) reading: "In nomad Arabia are lakes of immense size producing much bitumen which is gathered by the neighbouring tribes. This is not surprizing, because there are many quarries of hard bitumen there, when, therefore, a spring of water rushes through the bituminous land, it draws the bitumen with it, and passing outside, the water separates and deposits the bitumen."

But although these strata of limestone often contain up to 25% of bitumen, it is unlikely that they have been worked up in early times generally. For the sources of asphaltic bitumen in and along the eastern coast of the Dead Sea provided a far purer material in large quantities. Many classical writers refer to the Dead Sea or "Lacus Asphaltites" and to the "Bitumen Iudaïcum" obtained from it. From these reports it is clear that the most important seepages were situated, then as now, in the Dead Sea itself and that they yielded a viscous semi-liquid asphaltic bitumen, or as Pliny clearly calls it "slimy bitumen" (111). This is stated still more graphically by Diodor (112) who tells us that the Dead Sea: "Is a large sea which yields up much asphalt and from which a by no means negligible revenue is drawn. The sea is about 600 stadia in length and 60 stadia wide. The water stinks and is exceedingly bitter so that fish cannot live in it, nor do any other aquatic creatures occur in it. Although large rivers of very fresh water flow in it, it remains bitter. Every year a large quantity of asphalt in pieces of more than 3 plethra float in the middle of the sea, but often they are only 2 plethra in length. That is why the Barbarians who live on the shores of the sea call the large pieces "bull" and the smaller pieces "calf". When the asphalt floats in the middle of the water, it looks like an island to those standing on the shore. The advent of the asphalt is heralded 20 days before its arrival, for all around the lake the stench is wafted by the wind over many stadia and all the silver, gold and copper in the neighbourhood becomes tarnished; but the tarnish disappears again when the asphalt rises to the surface. The district in the vicinity, which is readily inflammable and which is pervaded by an unpleasant odour, makes the people's bodies ill and they die young."

Diodor has a similar statement (113) to make about a lake "in the land of the Nabataeans", the present Transjordania, which passage must be considered to be merely a duplication of the one quoted above.

It is a very good description of an oil seepage emitting hydrogen sulphide (H_2S)-containing gases. Similar phenomena are well known to occur at the present day in Mexico, Trinidad and many other parts of the world.

In a similar record, Josephus (114) says: "The change in the colour of the (Dead) Sea is a marvel. It changes three times a day if the rays of the sun enter it differently, they are reflected differently". This is probably an allusion to the interferential colours caused by a thin layer of oil on the surface of the water. It also transpires from the writings of Strabo (115) that: "It is full of asphalt. The asphalt is blown to the surface at irregular intervals from the midst of the deep, and with it rise bubbles, as though the water were boiling; and the surface of the lake, being convex, presents the appearance of a hill".

Here again Strabo gives details on the development of H_2S containing gases, which details we need not repeat here. The fame of the "lacus Asphaltites" lingers on and we find references to it in the writings of Suidas (116) and Isidore (117). Long after the bituminous materials had fallen into disuse the writers of the Middle Ages still recount the wonders of the Dead Sea.

Thus the author of *Voiage and Travayle of Sir John Maundeville* (a precursor, indeed, of Baron von Münchhausen) writes the following (Cap. XXX): "And two mile from Jericho is flom Jordan and you shall wete the Dead Sea departeth the land of Juda and of Araby and the water of the sea is right bitter and this water casteth out a thing called aspatum, as great pieces as a horse."

In the translation of the Bible by Luther we find that "The vale of Siddim was full of slime-pits" (118). The fact that "asphaltos" has been translated as "slime" (or "glue" in the Dutch *Statenbijbel*) is eloquent of the oblivion into which bitumens had fallen since Antiquity. The glance pitch deposits by the Dead Sea seem to have been neglected in Antiquity, these days they are mined for local use only.

We have some slight evidence on the exploitation of the Dead Sea bitumen. In this matter we must partly advance an opinion contrary to that of Lucas (119) and maintain that bitumen must have been of sufficient importance to Egypt to guide its foreign policy in a way at least in the Hellenistic period. We can read in Diodor (120) that Antigonus I, when at war with Ptolemy I, attempted to get the bitumen fishery on the Dead Sea into his own hands as a means of furthering the war. This, however, goes to show that it was very important to Ptolemy. Hieronymus, the governor of Idumea, failed to wrest the east side of the Dead Sea from the Arabs who at that time seem to control the fishery and to export the bitumen to Egypt.

That Egypt does not produce bitumen itself is shown by the fact that "asphaltos" is not among the Ptolemaic monopolies, though every other non-vegetable product of the earth was, e.g. salt, "natron", all mines and quarries. At some time during his reign Ptolemy II succeeds in taking the east coast of the Dead Sea from the Arabs, i.e. secured the important bitumen fishery for himself and Egypt. We do not know exactly what Ptolemy II did with the substances produced in foreign countries under his rule which, had they been produced in Egypt, would have been government monopolies. His control over this stretch of Nabataea amounted to a monopoly, as he controlled the largest source of economical supply of bitumen for export to Egypt.

At a later time the east coast was lost to the Seleucids, and then acquired by the Nabateans from whom Antony took it to give it to Cleopatra, which means that Egypt again got the control of the bitumen fishery. Cleopatra leases the sole exploitation to Malchus the Nabatean for 200 talents a year in 36 B.C. and when he fails to pay, Herod punishes him in 32-31 B.C. on the instigation of Antony and Cleopatra (121).

The further history of the bitumen fishery is again lost in the dark.

We will now turn to *Syria*. It is not definitely known whether the many deposits of bituminous limestone and the like were exploited in Antiquity. Very likely they were in Sûk-el-Chan, where it is said that open-cut mining was started about 1600 B.C. Large deposits of a very pure asphalt (containing at most $\frac{1}{2}\%$ of ash) are also found near Kfarje between Beirut and Ladikyeh, where even to-day great quantities are mined. (Table V). There are, unfortunately, few ancient accounts of development in this region.

Isidore (122) mentions oil seepages in Syria and Vitruvius (123) mentions "lakes producing bitumen" near Joppa, but does not give a detailed description of the source. Dioscorides mentions Sidon as the source (124), which, according to Engler, is a reference to "Siddim", hence to the Dead Sea.

This idea is incorrect, for Pliny refers to Sidon too (125) and adds that the bitumen occurs here "*as an earth*". There can, therefore, be no mistake in supposing that, not the Dead Sea, but Sidon and probably the bituminous limestone and may be, the glance pitch of Sûk-el-Chan is meant, though the latter must have been of little use in Antiquity foı several reasons which we have already discussed.

The only archaeological find of interest to us is a pendant from Byblos in which pearls and coloured precious stones are fixed in gold filigree with bitumen (126). Strangely enough, we know of no texts from these parts which make any mention of bitumen; not even the recently discovered Ras Shamra texts make any reference to it. A layer of material, identified by Claris (127) as bitumen, has been found at the bottom of a sarcophagus at Byblos, but the analysis is not very convincing. Séguin (128) felt justified in assuming that the Phoenicians traded in bitumen and took it even to Carthage, where it was used for embalming the dead. Without further proof, based on analyses, we feel that these assumptions should be treated with reserve, particularly as Séguin's data are largely derived from Mover's book on the Phoenicians which, in the light of more recent archaeological discoveries, has become somewhat antiquated. Nor is it an established fact that, as Moret (129) states, bodies in Canaan were wrapped in bitumen-coated mats and then burned on funeral piles, as they were in Babylon.

Along the *south coast* of *Asia Minor* several deposits of bituminous material were known, though as far as we can judge from their small size, they cannot have played an important role in ancient technology. In Cilicia there seem to have been small oil seepages, about which Vitruvius (130) reports: "At Soli, a town of Cilicia, there is a river named Liparis, and those who swim or wash in it, are oiled by the waters." Theophrastus (131) mentions "an earth from Cilicia, which becomes viscous on boiling" and Dioscorides (132) mentions the same "ampelitis" as coming from Seleucia. Similar descriptions are given by Oribasius (133), Aetius (134) and Galen (135), and according to present day knowledge small deposits of asphaltites occur here.

Strabo is probably referring to the same deposits when he compares ampelitis from the Pierian Seleucia to that of Rhodes (136).

Lycia is often mentioned as one of the sources of gagates, amongst others, by Dioscorides (137), Pliny (138), Solinus (139) and Isidore (140), the last two saying that it occurs more frequently in Britain. Bailey (141) makes its identification with some black kind of lignite very plausible. It figures largely in ancient medical texts and Epiphanios describes various applications (142). Smaller oil seepages seem to have existed in Lycia too, they are often mentioned and occur even in medieval literature (143).

Saint Nicolas, the famous bishop of Myra is often, during his lifetime, tempted by the Devil, who persuaded his parishioners to smuggle into the church an oil "that burned against nature in water and burned stones also"; the Saint always succeeded in exposing the "unnatural qualities" of this hellish gift. Was this crude oil or "Greek Fire" (144)?

Bituminous limestones or marls and rock asphalts occur in the former state of Armenia but do not seem to have been tapped in Antiquity.

Seepages of crude asphaltic oil near Samosata on the Euphrates were, however, well known, for we read in Pliny (145): "There is a lake at Samosata, a city of Commagene, which exudes an inflammable mud called maltha, it sticks to any solid body which it touches, and its power of adhesion enables it to pursue those who seek to flie from it, with this they defended their walls when besieged by Lucullus, and set his soldiery on fire, arms and all."

The Babylonian Talmud (146) mentions naphtha in Cappadocia, but gives no further details.

We now know, partly thanks to the geological exploration of *Mesopotamia* (Iraq) and *Iran* in the last few years, that numerous forms of deposits of bituminous products occur at the surface of the earth there. These deposits were also known in former times, and it is because the earliest inhabitants of Mesopotamia, the Sumerians, were keenly interested in the manifestations of nature that records of these phenomena have come down to us in the writings of their time.

The first zone lies between the Tigris and the slope of the Zagros Mts. (Map I). Our map shows numerous surface deposits, oil seepages, outcrops of bituminous limestones and gas wells. Unfortunately no analyses of these materials are known but asphaltic bitumens prepared from different Iraq crudes all show a relatively low sulphur content, thus furnishing a means of determing (partly) the origin of bitumen from ancient applications. This region is now explored and developed by the Iraq Petroleum Cy. (Table VI). The second zone is in Persia on the other side of the mountain range, the country on both sides of the Karun river, around the ancient Susa being the richest in oil. These fields are at present exploited by the Iranian Oil Expl. & Prod. Cy, from whose crudes asphaltic bitumens are prepared which, amongst other characteristics, show a still lower sulphur content. Unfortunately here too no analyses of surface deposits have been published.

The third zone is the country around Hit and Ramadi on the south bank of the Euphrates.

Every kind of surface deposit, as discussed in the foregoing lines, can be found here. It must have been the most important centre in Antiquity and even now it is an important factor for local production and industry though neglected by the international oil concerns. This bitumen has a very high sulphur content. (Table V) (147).

Although it is precisely in the countries between these three zones that bitumens were used for a great variety of purposes, we have only very little information from Sumerian, Assyrian or Babylonian texts as to the places where they were found. We must depend almost entirely on classical records for our research.

In North Assyria several seepages of bitumen are known in the district of Zakho, but no record of their exploitation in ancient times is known.

Much more important is the district of Qaijarah and Qu'ala Shargatt, where to this very day heavy crude oil and bitumen is won from a lot of seepages for local use. Andrae (148) surmises that the bitumen used in Assyrian buildings, etc. was won in this district and though he was not able to prove it, this is very plausible.

We find an allusion to this district in the following passage from Ammian (149) saying: "In Assyria there is asphalt in a lake called Sosingite, in the bed of which the Tigris is absorbed, which flows on underground to rise to the surface again at a great distance away.

Naphtha also occurs here; this material greatly ressembles asphalt and is very though and sticky. Even if a small bird alights on it, it is drawn down, it can no longer fly and it disappears into the depths. Once this kind of moisture begins to burn, human intelligence will find no other means of quenching it than by earth.

A big cleft will be seen in these districts from which mortally fatal fumes rise up, the heavy odour of which will kill any living creature coming within its reach."

G. Smith (150) alludes to the lumps of bitumen found in the Tigris and in pools in the neighbourhood of Hamman Ali. It is very probable that this bitumen was used by the Assyrians, for Prof. Campbell Thompson informed me that: "It would appear that there is a word A-GAR-GAR-DINGIR-(AN)-ÍD, dung of the river, parallel to KI-BIR-AN-ÍD, bank of the river = kibritu = sulphur". This bitumen was used in Assyrian medecine, the name "dung of the river" is an exact description of its appearance in seepages or pools.

This is what Jubair had to report about this region in his days (151): "In the morning of that day, the 22nd of Safar 580 (June 4th, 1184) we passed a village called al-Kayyâra, not far east from the Tigris river to the right of the road to Mosul. There we saw a black depression in the ground which ressembled a cloud. Here God has shaped a number of larger and smaller wells which cast forth kar (bitumen). Some cast out from time to time bubbles of that material as if they were water-pipes. Basins have been constructed there in which the material is collected which then looks like black, slippery, gleaming, moist clay, which is spread on the soil, which casts forth a pleasant (sic!) smell, coagulates strongly and adheres to the fingers at the first touch. Around the wells there is a large black pond on which something like moss floats which is driven towards the banks of the pond where it precipitates as bitumen. We thus saw something marvellous which we had heard of but which we believed untrue.

Not far from these wells on the banks of the Tigris there is another large well which we saw emitting smoke at a distance.

They told us that here a fire is lighted when the stuff has to be transported; the fire dries up the watery moistness and makes the material adhesive, after which operation pieces are cut out and transported. This material is traded down to Damascus, Acre and the entire (of the Mediterranean). God truly creates what he wishes. There is no doubt that a certain well, which according to our information is situated between al-Kûfa and al-Basra is just like the one we described above, that is why we mentioned it."

All along the Persian frontier in the neighbourhood of Baba Gurgur, Kirkuk, Kifri and Khasri Shirin many seepages exist. Even to-day the inhabitants of this district collect bitumen from the wells near Kifri at the foot of the Naft Dagh. An omen tablet suggesting burning wells near Kirkuk, says: "When in ditto a pit opens and IDDU BIL (= burning bitumen) appears..." (152).

These phenomena are well illustrated by a recently published photograph (153), and similar phenomena are alluded to in Isaiah XXXIV, 9, when an example of utter barrenness is suggested.

Then again many deposits of bituminous limestones occur in this region especially near Tuz-Khurmatli. To these, or perhaps to the seepages, the Sumerian patesi Gudea (2250 B.C.) is referring when he says that he "had the asphalt from the Magda mountains in Elam shipped (hence probably down the Adhem) to his city of Lagašh" (154).

The fact that so little bitumen was used in Assyria is remarkable, yet understandable. If we take the inscriptions left by various Assyrian kings we find the following:

Adad-Nirâri built his quay wall in Assur with bitumen in about 1300 B.C. (155). Upwards of four centuries later Tukulti-Ninurta II visited the seepages at Hit. In about 800 B.C. under King Adad Nirâri III, the governor, Bêl-harrân-bêl-uṣur, had many imprecations inscribed on his monument against any who should damage that work or cover it with bitumen. Sennacherib (156) built walls of limestone slabs with bitumen and Esarhaddon used brick and mastic for the foundations of his temple (157).

The preference for stone for building purposes in these districts and scarcity of suitable sources of bitumen in Assyria may easily account for its virtual absence. There is no mention of bitumen in the elaborate correspondence of the Assyrian merchants in Asia

Minor, in the Hittite empire. Nor, indeed, were any traces of bitumen found during the excavations carried out in that area (e.g. by the Oriental Institution at Alishar). Contrary to Séguin (158), experts inform us that there is no term for bitumen in the Hittite texts.

We will have occasion to discuss the difficulties of distilling crude oil or melting up rockasphalt in a country with the scarcity of fuel existing in Mesopotamia as far as we know, even in ancient times. We need not, therefore, be surprised to find that in Antiquity the greatest importance was attached to the third zône mentioned above, the region of Hit and Ramadi. For here seepages and streams yield bitumen in a form practically ready for use. The importance of this region is reflected in the many passages of classical authors referring to the sources of asphalt in Babylon. Vitruvius mentions an "asphalt lake" near Babylon (159); and Dioscorides (160), Pliny and Strabo refer to it.

Diodor (161) is very enthousiastic about this industry and says: "Whereas many incredible miracles occur in the Babylonian country, there is none such as the great quantity of asphalt found there. Indeed, there is so much of it that it is not only sufficient for so many and such large buildings, but the people who have gathered there collect large quantities of it, and although the multitude is without number, the yield, as with a rich well, remains inexhaustible."

Strabo recounts some very interesting details, which are worth while repeating here (162):

"Babylonia produces also great quantities of asphalt, concerning which Eratosthenes states, that the liquid kind, which is called naphtha, is found in Susis, but the dry kind, which can be solidified, in Babylonia; and there is a fountain of this latter asphalt near the Euphrates river; and that when this river is at its flood at the time of the melting of the snows, the fountain of asphalt is also filled and overflows into the river; and that there large clods of asphalt are formed which are suitable for buildings constructed of baked bricks, other writers say that the liquid kind also is found in Babylonia. Now writers state in particular the great usefulness of the dry kind in the construction of buildings, but they say also that boats are woven with reeds and, when plastered with asphalt, are impervious to water.

The liquid kind, which they call naphtha, is of a singular nature; for if it is brought near fire it catches the fire; and if you smear a body with it and bring it near to the fire, the body bursts into flame. It is impossible to quench these flames with water (for they burn more violently), unless a great amount is used, though they can be smothered and quenched with mud, vinegar, alum, and birdlime. It is said that Alexander, for an experiment, poured some naphtha on a boy in a bath and brought a lamp near him; and the boy, enveloped in flames, would have been burned to death if the bystanders had not, by pouring on him a very great quantity of water, prevailed over the fire and saved his life.

Poseidonius says of the springs of naphtha in Babylonia, that some send forth white naphtha and others black; and that some of the former consist of liquid sulphur (and it is these that attract the flames), whereas the others send forth black naphtha, liquid asphalt, which is burnt in lamps instead of oil."

One is inclined to conclude that Strabo or his informer had a more intimate knowledge of the materials for this passage very accurately described the distinction between the light-coloured inflammable crude oil with much lighter (petrol) fractions and the thick, black, pretty harmless asphaltic crude, which by evaporation of a small percentage of lighter fractions yields asphaltic bitumen.

The majority of the bitumen used in ancient Mesopotamia seems, however, to have come from the seepages on the south bank of the Euphrates. This is clear when one compares the sulphur content of the bitumen in the ancient mixes (Table IV) with those of samples from Hit and other seepages in the neighbourhood (Table V) and with those of asphaltic bitumens from Iraq and Persian crudes (Table VI). This bitumen is won from seepages between the rivulets Kubessah and Mohammedyeh, and originates from underground bituminous limestone strata. Several hundreds of these wells are known around Hit, Ain Ma' Moora, Ain el Maraj and Ramadi. About 25 miles farther south, where the seepages are far from the river and from habitation, the bitumen has accumulated, and we have the big deposits of Jebba and Abu Gir. This oldest material is very hard and contains up to ca. 20% of mineral matter consisting largely of windblown dust. This material must be considered as rather unsuitably hard for use in Antiquity.

The idea that bitumen may have been imported from Palestine or Syria (where some surface deposits are also characterised by a high sulphur content) may be dismissed as they largely consist of glance pitch, a brittle material with a high melting point, which even nowadays is used only for the preparation of varnishes and paints, but is quite unsuitable to be worked up for mortars or mastics.

Spielman (163) describes these "asphalt-wells" in the following way: "At the present time asphalt is collected near Hit as it comes to the earth's surface. Water rises with varying velocity, sometimes accompanied with so much gas, that the latter will burn after being ignited. In the water are "snakes" of asphalt, which collect together and are consolidated by the natives by hand-pressure into lumps, which are then thrown aside. After a very short time they become flattened owing to subsidence under their own weight. It is possible that the material was collected in the same manner in ancient times, because similar lumps of asphalt have been found at levels only slightly above the Flood Layer at Ur (3000 B.C.?)."

I was fortunate enough to receive from Sir Leonard Woolley such a sample from Ur, which showed the following analysis:

a) Bitumen content 41.5% by weight
Sulphur content of the bitumen = 9.2%
Diazo reaction (for tar compounds) ... negative
Vanadium and Nickel in ash ... positive

b) Remainder (58.5% by weight) composed
of 90% by vol. of vegetable matter (rushes) &
10% of mineral matter

Mineral matter: very fine (⅔ finer than 0.074 mms.) composed of 50% feldspar, 40% calcite and 10% biotite, probably windblown dust.

Another similar lump weighing 10 kg had the appearance of being composed of much smaller pieces extracted from the seepages and it seems that they were joined while kneading out the occluded water. It was discovered in the "Flood Layer" at Ur where it had remained for many ages left over from the building activities in the early days of that important town. The bitumen content was found to be 70%, and the sulphur of the bitumen 6.8%. This natural bitumen from Hit contained 30% fine wind-blown desert dust of the normal composition of the sands found in the deserts round Mesopotamia together with traces of reeds and rushes (164). This largely confirms Spielmann's assertions.

The lumps of crude asphalt are freed of the occluded water by hand-pressure and transported in baskets. To avoid sticking of the bitumen to the basket, the lumps of asphalt are rolled in the desert-sand or the inner side of the basket is coated with sand.

The crude asphalt lump therefore always contains some windblown desert sand in its surface layers before it is intentionally mixed with sand, etc. to prepare mastics. But we need not depend on analytical data only to prove the importance of Hit as a centre of the bitumen-industry in ancient Mesopotamia, for the texts yield some valuable evidence too.

Many references are made in the oldest literature to Hit (Akkad. Id = Greek Is) and Ramadi (Akkad. Rapiqu(?). Prof. Campbell Thompson says: "Iddu, the proper Assyrian word for bitumen, may have given (or taken) its name to (from) Hit". The ancient importance of Hit also emerges from the fact that cuneiform texts in the British Museum allude in more places than one to the rivergod Id, which was worshipped in the Babylonian pantheon (165).

In the times of Sargon the town must have been important enough for he came to offer tribute to the fish-god Dagan (the Dagon of the Old Testament) in his temple Ekisaga in Hit (2350 B.C.) (166).

According to Rawlinson (167) we find among the tribute of Mesopotamian cities, recorded by Tothmes III in his inscriptions at Karnak, "2040 minae (1020 kg) of sifte from the chief of Ist", to which he adds that "sifte" must mean the same as the Arabian zift = pitch, bitumen or incense.

Unfortunately this translation is due to a misinterpretation and it seems to have nothing whatever to do with Hit or with bitumen. The passage mentioned by Rawlinson occurs in the Annals of Tothmes III in the account of his ninth campaign (168) and refers to the tribute of the chief of "isy" (probably Cyprus?), which consists of copper in manifold form, e.g. the "sftw" copper perhaps meaning something like unrefined or "blister-copper" not yet cast in the form of bars for regular trading purposes.

Natural gases have always played a certain part in history, although their importance is sometimes exaggerated in technical treatises. These gas-wells were especially important in Iran because of the part played by fire in Iranian religion (169). They were particularly awe-inspiring because, as the old name, varishnak, implies, they needed no food (170). Down to classical times we find traces of the part played in religious cults by burning natural gases. They are, for instance, depicted as burning near Apollo's Shrine (171) on coins of Apollonia, near the present Selenizza (Albania).

Natural gases were known in Mesopotamia, but had nothing to do with religious cults; they are, however, mentioned, in the omen-literature, to which we shall revert presently. There is also a passage

(172) in the annals of King Tukulti Ninurta II (888—884 B.C.), describing gas-wells near Hit as follows: "Opposite Id (Hit), close to the sources of bitumen (kupru), I camped at the place where the voice of the gods issueth from the Ušmeta rocks."

These Ušmeta or Ussipta rocks are gypsum strata seamed with bitumen and sulphur deposits; the gases, mixed with bitumen and water, forcing their way through these gypsum strata to funnelshaped fissures in the earth's surface, make a roaring noise, which was taken for the voices of the gods in the underworld and, therefore, for an oracle (173). Similar deposits from which roaring gas issues are found in the surroundings of Kirkuk, where they are accordingly called kirkuk baba or abu geger, which means "the father of sound".

The fame of Hit also reached Herodotus, who says (174): "There is another city, called Is, eight days' journey from Babylon, where a little river flows, also named Is, a tributary stream of the river Euphrates; from the source of this river is rise with the waters many gouts of bitumen and from thence the bitumen was brought for the wall of Babylon".

These were the wells that supplied Babylonia (and Sumer) with the ideal mortar for its brick architecture.

The very frequent use made of bitumen in Neo-Babylonian times in the south is all the more striking, but logical. Many inscriptions by Nabopolasser, Nebuchadnezzar, Neriglissar and Nabonidus mention both iddû and kupru, (175), in quantities up to 15 tons for the applications to be mentioned later on. Up to Persian times (Cambyses) the texts cite large quantities of bitumen (176), although by that time its use was fast diminishing. As these later texts do not appear to contain any particulars of technical interest and are accessible to all who wish to study them in the publications cited, we shall not discuss them all here. There are, however, a few general statements that are of sufficient interest to be mentioned. One letter from the ancient town of Uruk (177) states that wine and bitumen will not go up the river in ships at the same time. Another letter shows that there is a local shortage of bitumen, and somebody writes, not without exaggeration: "Concerning $\frac{1}{2}$ še (23 mg) of bitumen for the fifth time I have written to you and you do not send it. Find me $\frac{1}{2}$ še and send it!" (178). Another letter complains in a similar strain (179).

When bitumen fell in disuse in the Persian and Hellenistic period, the seepages seem to have been exploited for local industries only.

Then for many centuries we hardly hear anything about them

until European travellers explore Mesopotamia again in the 16th century and tell us about a revival of the use of bitumen, which does not appear to have been recorded by native writers except Jubair.

One of the first is Cesar Fredericke, who in his voyage to the East Indies... (1536) says: "These barks of the Tigris have no pumps in them, because of the great abundance of pitch, which they have to pitch them withall; which pitch they have in abundance two days journey from Babylon. Near unto the river Euphrates, there is a city called Heit, nere unto which city there is a great plain full of pitch, very marvellous to beholde, and a thing almost incredible that out of a hole in the earth, which continuall smoake, this pitch is thrown with such a force that being hot it falleth like as it were sprinkled over all the plaine in such abundance that the plaine is always full of pitch. The Mores and the Arabians of that place say that that hole is the mouth of hell and in truth it is a thing very notable to be marked; and by this pitch the whole of the people have their benefit to pitch their barks."

A similar account is made in *The voyage of Mr. Ralph Fitch* (1585), to which is added: "The men of this country doe pitch their boats two or three inches thick on the outside." A few further details may be taken from Mr. John Eldred's account of his voyage (1583): "3 miles from the town of Heit there is a valley wherein there are many springs throwing out abundantly at great mouths a kind of black substance like unto tarre, which serveth to make staunch barkes and boats: every one of these springs makes a noise like unto a smiths forge in blowing and puffing out of this matter which never ceaseth night or day. This vale swalloweth up all the heavy things that come upon it."

Relatively unimportant as this region may be in modern times, in Antiquity great quantities must have been won here. This is proved by the numerous contracts dealing with the sale of bitumen or mastic found in Mesopotamian excavations. Those found in Tello give quantities ranging from 10 qa to 50 talents (9 kg to 1500 kg). Also the many applications in architecture testify of a well-developed bitumen-industry (180).

We must now turn to the third zône, the region on the banks of the Karun river, round about Susa. Herodotus is the first to relate that bitumen is obtained from oil wells near Arderica in the land of Cissia. This place Arderica was "210 stadia from Susa and the well was another 40 stadia away", hence in the neighbourhood of the modern townlet of Qirab.

Philostratos (181) gives the same information and adds that in this country "the soil is drenched with pitch and is bitter to plant in". Pliny (182) mentions oil wells in Susiana and in one of his other books (183) says: "Here flows the river Granis through Susiana, on the right bank of which the Deximontani dwell, who manipulate bitumen." This last quotation probably refers to the region around Bushire. For here and all along both shores of the Persian Gulf, notably at Koweyt, Bunder Abbas and on the island of Bahrein, some seepages and rich deposits of bituminous limestones and sandstones are known, and oil fields now produce large quantities of crude oil. The importance of this region seems to have been overshadowed in the Hellenistic age by *Media* especially the districts along the coast of the Caspian Sea and the river Araxes (around the present Baku), Plutarch mentions that Alexander the Great saw burning gas wells near Ecbatana, which he describes as "a gulf of fire, which streamed continually as from an inexhaustible source." The existence of seepages are mentioned too (184). About the much richer Araxes region there are very few notices. Pliny (185) mentions a good kind of pitch coming from Pontus where, however, hardly any products of this kind are known. This passage therefore probably refers to bitumen from Baku (or from Armenia?).

There are certain indications that crude oil and similar products were traded and imported in the Imperial age from the chief centres of production, Mesopotamia and Parthia. Thus Pliny mentions imports from Babylonia (186). A good deal must, however, have come from Media, as crude oil is referred to, even in Byzantine times, as "pir medikon", i.e. Medean or Median fire as two groups of ancient writers try to prove. One group derive the word from the famous Medea and maintain that the crude oil got this name by its inflammability or as Pliny says: "Naphtha is closely related to fire, which leaps upon it from any quarter as soon as it beholds it. With it they say Medea burned her husband's mistress, when fire seized upon her crown as she approached the altars to sacrifice" (187). Others maintain that oil from Media is meant and they are probably right. Crude oil called mydiacon (188) was certainly imported from the East in Byzantium round about 200 A.D. and could easily have been transported to that town by the Black Sea.

Again we hear that in 624 A.D. the emperor Heraclius invades Parthia with his army via Baku and in the North-Western oil area destroys many temples of fire worshippers bowing down before the burning natural gas wells. During the 10th century interest in petroleum products reawakened and sources were sought to supplement those already well known in Mesopotamia. And so we find Mas'udi (950 A.D.) writing about a land rich in wells of burning gas "Nefala-land" and relating how white and black naphtha are obtained there. A record by Ibn Hauqal (189) dating from the same period comments on the wealth of "naft" in Northern Persia "from the holy fire of which black soot is collected which is used as a dye or for the making of black writing ink."

It is only in the 13th century that the fame of Baku became known to Europe through the writings of Marco Polo (1272 A.D.) (190) who says: "To the North of Armenia lies Zorzania near the confines of which there is a fountain of oil, which discharges so great a quantity as to furnish loading for many camels. It is good for burning, in the neighbourhood no other is used in their lamps and people come from distant parts to produce it."

In North-east Iran Pliny (191) mentions wells of naphtha, viz. "among the Astaceni in Parthia" but is probably mistaken and means seepages by the river Oxus where in the present Turkoman Republic, U.S.S.R. drilling of oil-wells was started some years ago. Plutarch relates how these seepages were discovered by Proxenus a Macedonian, who had the charge of the equipage of Alexander the Great, when opening the ground by the river Oxus to pitch his masters' tent (192). The greasy oily liquor "became perfectly clear, when the surface was taken off, and neither in taste or smell differed from real (olive) oil nor was it inferior to it in smoothness and brightness, though there was no olivetree in that country!"

When Strabo reports that "it is said that people digging near the Ochus river found oil" (193), he seems to have misunderstood his informant, as he clearly shows in the same passage that he distinguishes between the rivers Ochus and Oxus. This source is mentioned many years later by Al Mukaddasi who (975 A.D.) states that "naft and qar" (bitumen) were discovered in Transoxania; it is then repeatedly mentioned and Alqazwini (1275 A.D.) goes so far as to say that this region constitutes the principal source of petroleum production for Persia, which seems very improbable (194).

The excavations at Mohenjo Daro, Harappa, and Nal revealed the existence of an *Indus civilisation* which seems to have known and used bituminous materials. We are able to prove by analysis of the bitumen from Mohenjo Daro (Table IV) that this was not imported from

Mesopotamia, but must have been won locally. There are several places which suggest themselves; notably, asphalt wells or rock-asphalts on the Basti river near Iskardo (Kashmir) and in the Seria mountains (Hazara district) or asphaltic crude oil from Khatan, Rawalpindi, Mogalkot or Gondawa. Yet even to-day analyses or further particulars of this region are very scarce. So it was in Antiquity too. True, Vitruvius (195) mentions "an Indian lake which in clear weather produces a great amount of oil" and Dioscorides (196) goes so far as to call Indian bitumen the best kind, but they tell us nothing about where it was found.

We must now turn to *Europe*, where a very old important source was in Zacynthus (now Zante) on the coast of Albania. Here even today a very pure soft bitumen containing much emulsified water is found in wells and pools. This place is described by Herodotus (197) Dioscorides (198), Vitruvius (199) and Aelian (200).

In the neighbourhood, but on the mainland, the existence of the "pissasphaltos" of Epidamnus (or Dyrrachium) in the land of the Apolloniates was known. Dioscorides (201) says: "In the vicinity of Epidamnos there is what is called pissasphaltos. It comes down from the Ceraunic mountains, is carried along by the force of the current and is deposited by the surf upon the banks of the river, where it forms into lumps. It smells of pitch mixed with asphalt." (hence the name pix-asphaltos). Strabo (202) gives us similar information: "In the country of the Apolloniates is a place called Nymphaeum. It is a rock that gives forth fire; beneath it flow springs of hot water and asphalt—probably because the clods of asphalt in the earth are burned by fire. And near by, on a hill, is a mine of asphalt."

This may be considered to refer to the rock asphalt deposits of Selenizza which are still exploited to-day. Modern drilling for oil in Albania was, however, stopped since trials proved it unprofitable (203). Aristotle refers to petroleum as fairly common in the Balkan peninsula when he says (184): "Thick dark and tough crude oils flow beside natural pitch and asphalt in Macedonia, Thrace and Illyria from the hot burning soil (smelling of sulphur and bitumen) and diffuse stinking, choking and sometimes deadly fumes."

He is mistaken, however, in the case of Thrace where modern geology has been unable to find any bituminous deposits, though he possibly refers to the Thracian "spinos" which Theophrastus (204) mentions and describes.

Next comes *Sicily* where the exploitation of the rock-asphalt of

Ragusa is still a profitable business. This place is not mentioned by ancient writers, but they all refer to the river Akragas near Acragantium (the present-day Agrigento) where seepages yielded a thick asphaltic crude oil.

Daubeny discovered several of these seepages near a hill called Macaluba. Dioscorides refers to the Sicilian oil as follows (205): "Bitumen is found in its liquid state near Acragantium in Sicily. It floats on the surface of the springs and is used in lamps instead of (olive) oil. Those who call it Sicilian oil are mistaken, for it is an established fact that it is a kind of liquid bitumen."

Pliny (206) makes a very similar remark and so does Aristotle (184) who adds "that it often has a distinct odour of cedar resin."

It is quite possible that Theophrastus is referring to the rockasphalt of Ragusa in the following passage: "The stone which occurs on the promontory Erineas (near Syracuse) smells like asphaltos and after burning looks like burned clay." (207).

Finally, we know from discoveries in the lake dwellings of Switzerland that the Neuchâtel (Val de Travers) rockasphalt was exploited in that period and used locally (208). No other knowledge has come down to us of places where bitumens were found in ancient times.

When summing up the evidence collected in this chapter we are struck by two facts.

First of all, the most important deposits of bitumens coincide with the Fertile Crescent and especially with the eastern part. We may thus expect an early use of these materials. Secondly the most important deposits were situated over the frontier of the Roman Empire.

This is probably one of the main reasons of the disuse into which they fell in later Antiquity, when tar and pitch produced in Macedonia, the Troad (Mt. Ida), Calabria and other wooded districts became much more prominent.

Collecting and refining bitumen

Unfortunately, little is known about the means of collecting and refining bitumens employed in ancient times. Its principal consumers in Antiquity, viz. Mesopotamia and India, have left no records on this subject. Once again, we depend for our information upon a few brief statements made by classical writers. These, however, only tell us about the collection of the material and nothing about refining methods. It was a very simple matter to collect the crude bitumen from the Dead Sea, for, as Tacitus says (209): "Those who make it their business to collect it, draw one end (of the floating bitumen) into their boats; the rest of the mass follows without toil or difficulty and continues loading the vessel, till the viscous substance is cut in two", and "This extraordinary substance, floating in heaps up and down the lake, is driven towards the shore, or easily drawn by the hand; and when the vapour that exhales from the land, or the heat of the sun, has sufficiently dried and hardened it, it is cut asunder by wedges or the stroke of a hatchett."

Strabo (210) complements this record as follows: "It floats, because of the nature of the water (highly saline). They reach the asphalt on rafts and chop it and carry off as much as they each can."

The method of winning asphaltic crude oil in Persia, as depicted by Herodotus (211) is quite different: "...To the well, whence men bring up asphalt and salt and oil. This is the manner of their doing it:—A windlass is used in the drawing with half a skin made fast to it in place of a bucket; therewith he that draws dips into the well, and then pours into a tank, whence what is drawn is poured into another tank, and goes three ways; the asphalt and the salt forthwith grow solid, the oil (petroleum) which the Persians call "rhadinace" is dark and evil smelling."

So the odour of Persian oil was commented on even in those remote times! From the foregoing passage we may also conclude that the original seepage was dug out to form a well, perhaps with timber-lined walls, as has been the practice of primitive oil workings all over the world.

The separation of asphalt and salt from the crude mentioned in this must mean a separation of asphalt plus dirt in emulsion form from the oil and salt water, as no instance is known of asphalt settling out from crude oils.

The collection was more primitive and more difficult in Zante. This is also described by Herodotus (212): "I myself saw pitch drawn from waters of a pool in Zachynthus. The pools there are many; the greatest of them is 70 feet long and broad, and two fathoms deep. Into this they drop a pole with a myrtle branch made fast to its end, and bring up the pitch on the myrtle, smelling like asphalt, and for the rest better than the pitch of Pieria. Then they pour it into a pit, that they have dug near the pool; and when much is collected there, they fill their vessels from the pit."

That is very similar to the way in which they went to work in

Sicily, where according to Pliny (213): "It also occurs as a rich oily liquid in Sicily where it contaminates the waters of a spring at Agrigentium. The local folk collect it by means of bunches of reeds, to which it adheres very readily..."

This primitive method was still used in the early Renaissance in Germany, for Agricola (214) reports: "Liquid bitumen sometimes floats in large quantities on the surface of wells, brooks and rivers and is collected with buckets or pots. Small quantities are collected

Fig. 2.
Crude-oil production and refining
(from Agricola, *De Re Metallica*, 1556).

by means of feathers, linen towels and the like. The bitumen easily adheres to these objects."

Primitive as the method may be, surprising results are achieved by it. Thus it is known that in the eighties more than 200 gallons a day were collected by natives near Surabaya (Java) for a small local refinery by the very same means. Even if bitumen was obtained in Mesopotamia, Persia or Palestine in as simple a way as mentioned above, the crude product must have undergone further treatment, because it would not generally have been of the required hardness or consistency. This hardening was probably achieved by "drying or evaporating" in the air, a process perhaps indicated in Pliny's remark (215) that Judaean or Sidonian bitumen "can be thickened or condensed."

It is unknown whether this process was accelerated by gentle heating, which must have been used for the refining of rock asphalt or the preparation of mastic, but it is very probable that heat was employed in refining.

One difficulty stood in the way of using lighter asphaltic crude oils, viz. the lack of knowledge of the technique of distilling, which is necessary to remove the dangerous light fractions.

By distillation we mean evaporation of the lighter fractions by heating, and subsequent condensation of the gases by cooling (air or water) outside the distillation vessel. In archaeological literature no trace can be found of early distillation apparatus.

The earliest mention of distillation is often said to be a passage of Aristotle's *Meteorologica* (II, 3) on the formation of mist and rain and the formation of sweet water when the vapour of salt water condenses, though distillation is not actually mentioned. Still a form of distillation was known to Theophrastus when he described the production of wood tar, he did not recognise it as such.

True, there is mention in Hippocratic records of about 350 B.C. of the distillation of liquids in a "calabash" from which the vapours were discharged through a pipe sealed with loam but the evaporated fraction was not condensed and collected, the original liquid being merely partly evaporated to obtain the residue.

It may be, however, that the first trials in the art of distilling were made at about this time.

Still, Dioscorides, when speaking of "separating the liquid parts of the bitumen by distillation", does not mention the condensing of the vapours formed. At that time condensation methods were, however, occasionally applied, but naturally very primitive means were used. Thus Pliny (216) says that tar oil can be obtained by stretching a hide over a cauldron containing boiling pitch and then wringing out the condensed liquid.

Again, when Dioscorides wished to make lamp black, he burnt resin or pitch under a receptacle, which was cooled with wet sponges (217).

A generation later, however, i.e. about 100 A.D., we learn from the writings of Coptic alchemists, such as Pamnenes, Maria the Jewess and Cleopatra, that a distillation method was then known in which the vapour was delivered to a receptacle cooled with air or sponges. As was the way with many inventions, however, years elapsed before methods as these were applied on any practical scale.

In about the year 300 A.D. Zosimos had more elaborate theories about distillation and the sublimation of various substances in glass apparatus sealed with mastic and earthenware stoppers, and his methods were apparently adopted by Synesios in the production of volatile substances from various plants and "soils". After Synesios, however, we hear nothing more about this technique of distilling until round about 1000 A.D., at which period it is quite certain, that distillation methods were in use on a larger scale and were certainly being applied to petroleum. Not only had "Greek Fire" been in use since 700 A.D. and for which, besides crude petroleum, its light fractions are often specially mentioned as an ingredient, but frequent allusions to these light fractions prove that distillation became more common.

Though Ali Ben Abbas still uses Pliny's method of condensing in a hide (950 A.D.), his contemporary Abu Mansur discusses the distillation of water, calling the distilled water "arâq" i.e. sweat.

But two centuries later many references are made to the distillation of bituminous materials.

Alnabarawi (1200 A.D.) describes the distillation (taqṭir) of tar, while his contemporary al-Kazwînî (218) knows two kinds of nafta viz., white and black, the latter, becoming white by distillation with "helmet" (column) and "alembic" (hood). Similar remarks occur in the writings of Ibn al-Beitar and Ibn-al-'Awwân. Very full details are given by Dimashqi who tells us that the art of distilling oil is a well developed industry in Damascus and surroundings.

We need not go into details on this industry but may be allowed to close our survey by stating that distillation did not become known in the West until about 1250. Thenceforward a number of improvements follow in quick succession. In about 1300 the retort was introduced from the East and at approximately the same time Alderotti Florentinus mentions the tubular cooler, though still without continuous water-cooling. Even in about 1400 Michael Savonarola only knows of cooling the alembic with wet cloths.

The survey of this development, together with the lack of excavated apparatus convince us that distillation was not applied to crude oils in Antiquity (219).

I have mentioned a more primitive way of removing the lighter fractions of the crude oil by "drying or evaporating" in the air. Though no records of this method exist from Antiquity, we meet it in later times in those districts where petroleum was worked for local use only and not for export. Thus in about 1550 A.D. it was used in Germany, for Agricola says (220): "The liquid bitumen is collected in big copper or iron vessels and the lighter fractions evaporated by heating" and also (221): "The Germans of eastern Hungary and the Saxons usually prepare bitumen by heating the crude oil in copper or iron vessels. During this operation the crude oil was often ignited and the fire could be extinguished with wet cloths."

Sometimes the reverse operation, viz. softening the bitumen with thin oils, or as we call it at present "fluxing" or "cutting back" seems to have been practised too. Pliny tells us that "the test applied to

Fig. 3.
Smelting bitumen from bituminous rocks.
"Destillatio per descensorium"
(from Agricola, *De Re Metallica*, 1556).

ampelitis is that when it is mixed with oil, it should liquify like wax..." (222).

Olive oil seems to have been used for this purpose for Strabo states (223) that: "ampelitis was also discovered in Rhodes but it required more olive oil (than the Pierian)."

Another method of preparing bitumen remains to be discussed, viz. the working up of rock asphalt and similar natural products. Sometimes these rock asphalts contain so much bitumen that they can be used as such. This was certainly done in Antiquity for the Morgan found in Susa vases of asphalt which contained about 75% of filler (or fine mineral matter), wholly similar to the rock asphalts found in present-day Iran. These fillers are too fine for artificial incorporation therefore a natural product must have been used.

Now we have described many deposits of rock asphalt but as a general rule they contain less than 20% of bitumen, the average rock asphalt about 5—12%. Bitumen can be made from rock asphalt by selecting the darker parts of the deposits (richer in bitumen) and melting them down, or if this is impossible on account of the low bitumen content, liquifying (fluxing) them by the addition of a quantity of pure bitumen.

In both cases a portion of the mineral matter, viz. the coarser particles and other adulterations, will settle out from the molten mass, while the finer mineral matter (the so called "filler") is kept in suspension.

The presence of this filler is very characteristic for bitumen obtained from rock asphalt for it is only possible to incorporate them artificially with the aid of modern machinery. If we therefore find such fillers (which are much finer than cement particles) in ancient bitumen samples, we may be sure that the bitumen used was a purified rock asphalt or the like.

This simple method described above is the principle underlying modern refining methods at Trinidad, Neuchâtel, etc. The refined rock asphalt or "épuré" which is marketed still contains an appreciable amount of filler (in the case of Trinidad 45%) and lacks the brilliant black lustre of asphaltic bitumen from crude oils. A variant of this method, more suitable for rock asphalts with a low bitumen content must have been known in Antiquity too, for it is referred to by Aetius (540 A.D.) as a matter of common knowledge. This is the method called "destillatio per descensorium" in the Middle Ages. This process was carried out in two super-imposed jars separated by a screen. A fire was placed around the top jar filled with the material under treatment, and the bitumen that was formed dripped through the screen into the bottom jar embedded in the soil. The distillate (lit. "drippings") thus prepared proceeded from the combined action of distillation, smelting and cracking (destructive distillation by overheating) of the bitumen in the rock asphalt, it was of inferior quality when compared with the bitumen prepared by the more careful method described above.

Ma'sudi (950 A.D.) uses it to prepare "oleum de gagatis", "gagates" meaning rock asphalt in this case. Many centuries later it was known to Agricola (220) who says: "Rocks which contain bitumen are treated in the same way as those which contain sulphur, by heating them in vessels with a sieve bottom. This, however, is not the common practice because the bitumen prepared in this way is not very valuable."

It was applied in Seefeld (Tyrol) for the production of ichtyol (Dirschöl) from a kind of bituminous shale between the years 1576 and 1840.

Though no text in ancient literature can be connected with the refining of rock asphalt with certainty, it may be that Pliny is referring to the production of épuré, when he says: "The best part is that which floats on the surface when it (the pissasphaltos) boils." (224).

These methods of preparing bitumen from rock asphalt were much more expensive than those from seepages or from asphaltic crude oil especially in a country poor in fuel. In as timberless a region as Mesopotamia it must have been uncommon for bitumen to be produced in this way; analyses show that ancient samples lack the characteristic fine mineral matter of rock asphalt or refined products derived from them (Table IV). A mastic from Mohenjo Daro (Indus valley) is probably the only sample that was a refined rock asphalt.

Wood tar and wood-tar pitch were never able to compete with bitumen of which there was plenty in the East. Another hindrance to the increasing use of pitch was the ever diminishing stock of wood. The shrinking forests of these regions could not be wasted for this use. Pitch, therefore, could only become a serious competitor in regions with a practically unlimited supply of wood, such as some districts of Asia Minor and the Balkan Peninsula. Thus in the days of Hellenism Macedonia and Mt. Ida in the Troad competed on the pitch market of Delos. Glotz (225) shows in an interesting article how intimately the prices for pitch in Delian contracts are connected with the political situation in Macedonia, where the export of pitch seems to have been a government monopoly. At the same time we see that the high price, which may have been artificially raised by the monopoly, must have prohibited as general a use as that of bitumen in Mesopotamia.

In Caesar's time the production of pitch and tar was already a well established industry; quite elaborate descriptions of it have come down to us from Theophrastus (226) and Pliny (227).

The "boiling of pitch" consisted in stacking a large pile of wood blocks, covering it with a layer of earth or sod and then setting light to the wood. The tar produced by this form of "dry distillation" was drawn off through a drain leading from under the stack. According to Pliny this drain had to be sixteen ells long. Pliny further says that a better method is to dry-distil the wood in kilns ("furni" or "alvei") (228). It was, moreover, already known, that certain trees such as pines, spruces, cypresses and the terebinth were especially rich in tar (229).

In Italy the Bruttii became specially renowned (230) for their pitch, for their country Calabria had large woods of pine and spruce. Other pitch producing countries were Turdetania (231), the Alps (232), especially rich in tar (233).

The technique there employed in the production of tar gradually spread northwards and the wood tars of later fame, those of the Black Forest, Sweden and Norway undoubtedly owe their existence to the art of the Bruttii and Macedonians passed on to the more northern regions as the stock of suitable wood in the south decreased (234).

We are also told about the uses to which tar and pitch were put in the classical period. To mention only a few: waterproofing pottery (235), caulking ships (236), as a paint for roofs and walls (237), for the production of lamp black a base material for paints or ink (238).

Then, torches were made of it by soaking branches or oakum in pitch instead of using the more primitive bundles of pine chips (239). Finally pitch was used for modelling or as a core of hollow statues (240).

What interests us more, however, is the development and history of the use of bituminous mastic; but, unfortunately, little positive knowledge is available. In most ancient periods the mortar was undoubtedly often a mixture of loam and chopped straw or reeds, bitumen, and filler and fibrous materials (Table IV) in a composition remaining unchanged for centuries, with only slight variations for certain applications of different natures, such as mortar (35%) and asphalt mastic (25% of bitumen). This mastic appears so early in Tello, Ur and Uruk that Watelin said: "The use of bitumen is characteristic of Antiquity" (241). This mastic was, moreover, used in Tello as sealing-wax, and statues were cut out of it (242). Curiously enough, the mastic from Susa in Persia, which Berthelot analysed, has the same bitumen content (28%) (243). The bituminous mortar or mastic then remains practically unchanged until the later Neo-Babylonian age when, particularly under King Nabonidus, almost pure bitumen was used for mortar (244). In Persian times the use of bitumen in all its forms fell completely into the background except for caulking ships. This has been proved again by Mercier, who published an excellent analysis of the mortar of the palace at Ctesiphon (245), a mortar often, but erroneously, said to contain bitumen. A short time ago Thureau-Dangin published the translation of an interesting clay tablet (246) on the use of mastic, in which is calculated the amount of bitumen required for the bituminous mastic coating of a floor of a certain area.

Finally, King Hammurabi's code of law (ca. 1780 B.C.) gives an interesting comparison between the cost of coating with bituminous mastic and that of other work. Par. 228 of that code states that a house may cost 2 shekels of silver per Sar (i.e. 35.3 sq.m) base, a sum equal to what may be charged, according to Par. 234, for caulking with mastic a boat of 7½ cub. m. capacity.

How great these quantities were is not known but contracts from Tello show that parcels from 600—1500 kg are commonly sold.

Some prices recorded by de Genouillac show that the mastic, prepared as we shall see by mixing the pure bitumen with mineral and vegetable matter, was about twice as expensive as the pure bitumen, probably because of the cost of the fuel employed for the mixing and melting. This is quite different nowadays when one pays about £ 15.— per ton for asphaltic bitumen and about £ 7/10/— per ton for mastic, because of the cheap filler added, mixing costs amounting to a small fraction of the price only. This is also proved by the facts that the minority of the contracts found at Tello are for mastic and small quantities only, the larger part calling for the pure bitumen. Thus the practice seems to have been to prepare the mastic on the spot and save the fuel for remelting it.

That the production of seepages was fairly great is also shown by the high rent of the Dead Sea bitumen fishery already mentioned. Still bitumen prices in Mesopotamia were not very high; if index figures may be trusted, the price of pure bitumen in the time of Gimil-Sin (c. 2000 B.C.) would amount to something like £ 19/10/— per ton, an amount surprisingly like its modern equivalent.

It would be very interesting to compare prices from other periods with those mentioned here if they can still be found in unpublished cuneiform material or in publications which the author has overlooked.

Though we know nothing about the amount of bitumen or oil produced per year we may infer from later reports that these quantities may have amounted to several hundreds of tons. From Arabian annals we know that 10.000 of oil pitchers and the same number of nafta were lost in a fire in the residency at Cairo (1077 A.D.), a quantity of about 100 tons of oil at a conservative estimate. Now Arabian methods of collecting were not much better at that time than those of the ancient Mesopotamians and the widespread use of bitumen in ancient architecture may go to prove that the amounts collected were certainly fairly large.

As was almost inevitable, the origins of bitumen have been debated since ancient times. We discussed the modern point of view in an earlier chapter and with this in mind we might be tempted to read a statement of the animal origin of petroleum into the saying of Philostratus that a worm appears in the river Hyphasis in India "which when melted down yielded a fat or oil" with a strong tendency to catch fire inextinguishable by water (247). But we are mistaken for the generally current opinion in Antiquity was that bitumens were formed from earth under the influence of fire. Strabo propounds this theory (248): "Asphalt is a clod of earth, which at first is liquified by heat and is blown to the surface (of the Dead Sea) and spreads out; then again by reason of the cold water, the kind of water the lake in question has, it changes to a firm, solidified substance and, therefore, requires cutting and chopping. It is reasonable, that this behaviour should occur in the middle of the lake, because the source of the fire and also the greater part of the asphalt is at the middle of it; but the bubbling up is irregular because the movement of the fire, like that of many other subterranean blasts, follows no order known to us. Such, also, are the phenomena at Apollonia in Epeirotis".

Strabo seems to have adopted the theory of Poseidonius, for he also says (249): "...Since as Poseidonius says, the earth that is thrown in the trenches (at Apollonia) changes to asphalt..."

Probably Pliny has the same opinion when he compares the solid and liquid bitumen with "earth" and "mud" (250). This theory is reiterated by Isidore (251), who states that bitumen is of a fiery nature; though he has only rather vague ideas about these materials generally. On the whole, of course, the bitumens were judged according to their physical and visible characteristics. We have seen some examples in Sumerian nomenclature, but some more details can be found in classical literature.

This is Pliny's opinion on the qualities which a bitumen should possess (252): "The excellence of bitumen is estimated by the height of its lustre, its density, and its heavy odour. When quite black it has little lustre, being adulterated with vegetable pitch. Its general properties ressemble those of sulphur..."

He, therefore, disapproves of the addition of pitch and looks upon it as a disadvantage, on which point Dioscorides agrees with him (253) for he says: "Indian bitumen is valued more than any other; it is considered so excellent because it glistens like unto purple, is heavy and emits a strong odour. The dark and dirty kind, on the other hand, is full of defects; it is mixed with pitch and comes from Phoenicia, Babylon, Zacynthus and Sidon."

There were several other properties of bitumen which had also been noticed in former times, such as the ductility. For Pliny assures us (254): "The bitumen, which is elastic and 'lazy', cannot be torn to pieces. It sticks to everything with which it comes into contact. It is then just as if a thread, bedded in the sticky asphalt, were stretched between that object and the bitumen." The adhesive properties of bituminous materials or their "stickiness" is mentioned in the Bible (255).

Tacitus tells us something similar and goes on to say that older historians declare that bitumen cannot be cut with iron or bronze, but only with a cloth soaked in blood. He knows, however, from his own experience that bitumen can very easily be cut into pieces if it only be allowed to dry and harden (256). According to him, this hardening may be accelerated by sprinkling in with vinegar "which provides the parts with the desired cohesion."

Pliny records the same hardening in the case of pitch (257) while Strabo handed down to us (258) an analogous observation made by Poseidonius, who obtains the hardening by using urine.

The statement that these substances harden bitumen and thereby make it brittle is incorrect. But the observation is justified as these substances, containing organic acids, will moisten the metal of hatchet or knife, and thus prevent the bitumen from adhering to it and facilitating the hacking up of a lump of bitumen.

The mumiyâ, or so called "bitumen", scraped from mummy linen or taken from the interior of mummies was of course nearly always a mixture of bitumen with other preservatives, etc.

This mumiyâ trade arose in post-classical times, but we must refer readers to the publications of Budge, etc. (259) for further details.

I will have occasion to prove that bitumen was only used in one period of Egyptian history for mummification purposes and that it is therefore wrong generally to regard the mumiyâ as bituminous material. Though wrong ideas on this subject are very common, modern analysis of these materials has given us sure proof of their heterogeneous nature.

Applications of bitumen

1. *Ancient mastics: their composition and preparation*

Before we proceed to review the most important uses of bitumen in ancient times we must answer the question: In which form was the bitumen applied?

The majority of the uses of bitumen requires a product which, while plastic, should not flow in the heat of the sun and in the majority of the cases pure bitumen was therefore excluded from use or else it would give a result which would be only partly satisfactory. Though pure bitumen was an easy material to handle, melt, etc. it was necessary to "stiffen" it by incorporation of fillers, fibrous materials or the like, in short as we call it, to make a *mastic*.

When a rock asphalt is refined, a product is obtained which has already some properties of a good mastic. The very fine mineral matter remaining in the épuré acts as a stiffening agent, giving it sufficient body to allow of its being plastered or trowelled. Such épuré was used in the Great Bath at Mohenjo Daro (260).

As I have already pointed out, however, that the pure asphalt of Mesopotamia was for several reasons the most important source for the bitumen supply and this very pure asphalt had to be mixed with sand, gravel, filler, etc. to obtain a good mastic.

I must needs limit my discussion on the materials used in ancient mastics, for Table IV shows all the modern analyses available of the mastics themselves.

It will be seen that nothing is known about the composition of bituminous mixes outside of Mesopotamia in ancient times. We must therefore limit discussion to this region and first of all try to find out, what kind of mineral matter was used.

The identification of *mineral matter in old mastics* is undoubtedly the most difficult part of research work of this kind. Meanwhile for one region, the number of possibilities can be limited to a large extent by a process of elimination, even if the minerals can not always be established with certainty.

In the first place it is evident from Table IV that no coarse materials such as gravel have ever been used in any of the samples analysed, which again must be a result of the scarcity of such materials in alluvial Mesopotamia. The Babylonian mortar (sample L) was the only one in which Dr. Herrmann found coarser material, namely fragments of baked bricks (5—15 mm large), bedded in a mortar of the recorded composition. There are also chips of brick in the samples A and K from Tell Asmar, but they are never larger than about 3 mm.

DESCRIPTION OF THE SAMPLES MENTIONED IN TABLE IV[1])

A. Mortar of brickwork found at Tell Asmar, dating from the Jemdet-Nasr period (2800—2600 B.C.).
B. Piece of mastic taken from the floor of a bathroom in a temple at Tell Asmar (Proto-Dynastic period, 2600—2500 B.C.).
C. Mortar of brickwork Tell Asmar (Early Dynastic, 2400 B.C.).
D. Mastic used as insulating layer on kiln in Tell Asmar (Early Dynastic, 2400 B.C.).
E. Mastic of floor in bathroom of a private house. Tell Asmar (Akkadian period, 2350—2150 B.C.).
F. Jointing and mastic of a floor, Tell Asmar (period Third Dynasty of Ur, 2050—1950 B.C.).
G. Mortar from brickwork, Tell Asmar (Larsa period, contemporaneous with the first King of the First Dynasty of Babylon, 2000 B.C.).
H. Mastic layer of threshold, Tell Asmar (Beginning of Isin-Larsa period, one or two generations after Third Dynasty of Ur, Nur-akhum, 1900 B.C.).
K. Mastic layer found on steps, Tell Asmar (Larsa period, Ibiq-Adad I, 1800 B.C.).
L. Mortar taken from Neo-Babylonian floor of one of the inner courts of Nebuchadnezzar's palace in Babylon. (Not further described; derived from excavations made by the Deutsche Orient Gesellschaft there, handed to Dr. Temme for examination by Dr. Wetzel, analysed by Dr. P. Hermann) (605—561 B.C.).
M. Mortar in water-coursing layer in wall of the Great bath at Mohenjo Daro (Indus Valley), taken in 1933 (2500—2200 B.C.).
Details on the method followed in analysing the samples are given in a previous publication (261).

These old samples are, therefore, built up of practically the same materials as the bituminous mastics of to-day, that is, besides bitumen, of a sand and a fine filler.

However, the discontinuous grading of the mineral matter (Table IV) is not caused by the use of two materials of which one represents the "sand" and the other the "filler". For the chemical analysis of these supposed components are so similar, that we must conclude that the discontinuous grading was, at least partly, caused by the method of crushing. A striking factor in all the samples is the large percentage

[1]) The above letters will be used in subsequent pages to represent the samples against which they stand.

of lime (CaO) besides the quartz (SiO_2) content. This considerably limits the number of possible mineral compounds, namely to: 1) loam; 2) limestone or marl, eventually chalk; 3) lime; 4) pulverised brick.

1) Generally speaking loam or clay cannot be identified chemically, and it is even difficult to classify soils owing to the prevailing confusion as regards the definition and nomenclature of loam, clay, loess, etc. In our case, however, chemical identification is possible, for Sir W. Willocks (262) explicitely states that "the characteristic soil of Mesopotamia is a light calcareous loam, unusually rich in lime". Their typical feature is an amount of about 12% (wt) of lime and about 48% (wt) of insoluble matter and quartz; indeed a proportion of lime to quartz, which will not be found in any other soil in the world.

Another inference from this high lime content is the unlikelyhood of all the loam in Mesopotamia being a fluvial deposit although some, indeed, have referred to important stretches of loess in that country (263). Loess is a windblown product—an aeolian deposit—its granular size being very characteristic, namely the main part, 50—60%, comes between 0.074 and 0.010 mm and containing little or no material between 0.010 and 0 mm, which is a very typical granular size for common clays.

It may therefore be regarded as a significant fact that in all the samples examined the typical clay gradings form a very small proportion.

In view of the above mentioned unusual lime/quartz proportion of Mesopotamian soils, it may be safely concluded that in the samples A, C, G and H loam has been added to the bitumen.

Moreover, it is obvious that loam would be put to this use in these regions for, not only was it the raw material for dried and baked bricks, but it was also used as mortar in the brickwork of the earliest periods, whilst it served as plaster on thatch or brick-work.

Now the strength of loam mortar after drying for three weeks is about 5 kg per square centimeter, which is sufficient for low buildings, but it can be greatly strengthened by the judicions addition of sand, straw or chopped straw, as it then has less tendency to crack, when dried out.

I consider it highly probable that there was an evolutionary process in which the loam mortar became gradually replaced by a bituminous mortar which contained loam, but to which other additions were made, which had proved to be useful in the case of loam mortar.

These additions served their purpose when applied to bituminous mortars although it was for different reasons of which the Sumerians may only have had only a vague idea.

2) This reasoning leads me to the consideration of the possibility of limestone, marl or chalk as a mineral compound used in mastics. As will be shown, I have reasons to suppose from own analysis that these materials were used for this purpose either pulverized or in the form of quarry dust. Then of course limestone and marl were very ordinary building materials where quarry stones were used and are to be found in many varieties in the areas bordering on Mesopotamia, whence they were imported of old.

3) The possible use of lime must also be discussed. The secret of preparing lime was known in early times, but the use of lime as mortar seems scarcely to have become popular until the Babylonian period (264), at that time it was in more frequent use and it is in that period that we hear of bitumen-lime mixtures as mortar. Banks, however, records it in Bismaya (265) and Campbell Thompson states that it is used more frequently in Assyria as mortar (266), then of course it is very often used in stucco (267). An exception to this was Assyria where sand-lime mortars were always popular even with baked bricks (268) though bitumen is often used in the waterproofing of bathrooms, drains and the lower two feet of walls.

If present-day custom may be considered to reflect a still living tradition, the following observation made by Parkhurst (269) is interesting: "As employed in repairing Baghdad roads, the almost pure bitumen or seyali is mixed hot with lime and mineral aggregate until it resembles a mastic after which it is poured on the road surface and worked into place with a small hand roller."

This use was confirmed to me by Mr. F. L. Bassett (Chief Chemist to the Government of Iraq, Baghdad), who added that "sand" (viz. the impure calcareous desert-sand) is generally used besides lime. It is not impossible, therefore, that lime was also used in ancient times as a filler but it is difficult to identify chemically, because it was in course of time, without doubt partly, if not wholly, converted to carbonate so that it chemically no longer differs from limestone. As for the older Tell Asmar samples microscopical examination and the chemical analysis (exactly corresponding lime and CO_2 amounts) would point to the use of limestone and the like, which as we know, were formerly also used in loam plasters as thinners.

It would seem then that the use of lime in bituminous mixes became more common in the Neo-Babylonian period; but this hypothesis ought to be confirmed by the examination of fresh samples.

4) A priori the use of pulverised bricks might also be assumed if analogies may be drawn from prevailing custom in these countries.

Helfritz (270) says that moist loam is kneaded with marl, fragments of bricks and fine straw to make new tiles or mortar even at the present times. Now we have tried to prove this in the Tell Asmar samples by using Goldschmidt's method, as described in various publications (271), which is briefly as follows.

He makes thin sections of the mineral matter after heating the sample (500—600° C) and claims that by this method and microscopial observation an opinion may be formed of the nature of the clay or loam used and also of its preliminary handling (kneading totally obscuring the original orientation of the clay minerals).

Our Tell Asmar minerals treated according to this method showed a totally different picture from a section of an Akkadian brick. Thus my opinion is confirmed that no pulverised brick was used in these mastics as Dr. Herrmann supposed, so that the minerals used are limited to loam, limestone or marl and perhaps lime (Table IV).

Many of the ancient mastics have been shown to contain *vegetable matter* and it is well known that straw and similar material is frequently used in Mesopotamia too in strengthening loam mortar as according to Helfritz (270) the mortar is often prepared to this day from fragments of sun-dried brick and gypsum (or a mixture of chalk and ash) and strengthened by the incorporation of reed matting. The similarity to our mastics is certainly striking and it again leads me to suppose that the mastic was evolved from the loam mortar and that it profited by the experience gained by the Sumerian builders with the latter.

Of course, reed mats, "burû", had in the past been used as a material for walls when they were coated with loam or bitumen. Later on these mats were used in masonry as a reinforcement, sometimes as loose layers of reed or rushes placed diagonally. Ropes made of rushes used as staying and stiffening material in brickwork have remained preserved even without the protection of bitumen, as Jordan discovered in the course of his Erech excavations.

Meanwhile, choppings of reed or straw in combination with loam had also come into use, an application borne out by later writings in which the use of this loam with straw is recorded for the making of bricks (272).

The reason why did the Sumerians resort to the addition of fibrous material to their bituminous mastics was that they could obtain similar advantages to those accruing from the addition of fibrous

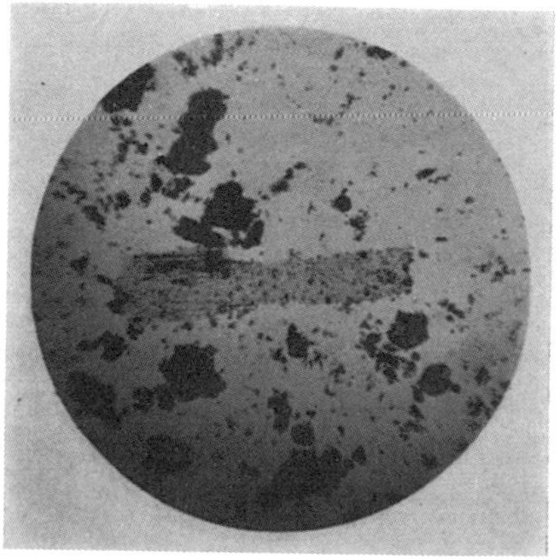

Epidermis of a species of Granimeae (from Sample C).

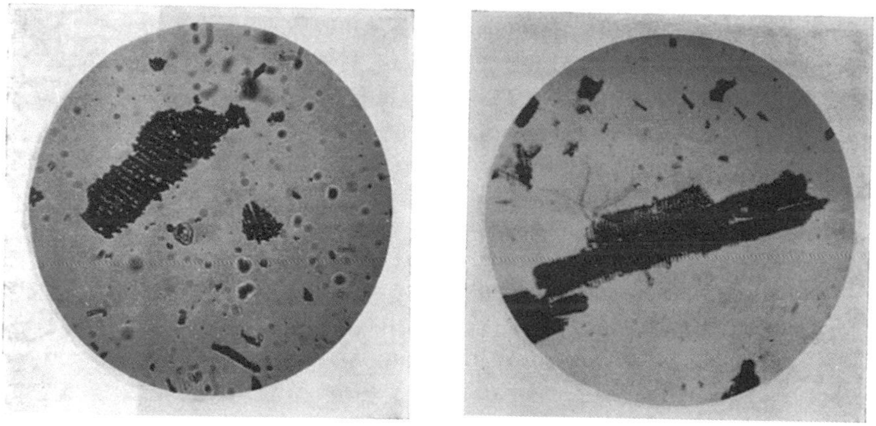

Two photographs of remains of reeds, distinct vascular bundles (from samples E & G).

Fig. 4.
Vegetable remains from ancient mastics.

material to loam mortars with this cheap material. They were able to impart sufficient rigidity to the mastic and saving them from the consequences of using a compound with a large excess of bitumen, and therefore easily softened by the sun or damaged by heavy loads, for these ancient mastics all show a great excess of bitumen.

When mixing mineral matter with bitumen, the latter upto a certain percentage acts as a glue in holding the grains together and forming a solid mass. If one raises the bitumen content the grains no longer adsorb the bitumen and this excess acts as a lubricant, causing sweating through heat or flowing under pressure, even at room temperature, if the excess were great. Now this can easily be prevented by the incorporation of a small percentage of fibrous materials and this was probably the underlying intention in olden times, for the slight disadvantage of a little less fluidity in the heat is more than compensated by greater stability after cooling to room temperature.

Modern asphalt technique avails itself again of these properties of fibrous fillers.

Judging by the Assyrian texts translated by Campbell Thompson (273) the following kinds of reed, rush or cereal growing in Mesopotamia might have been used as base material for the fibrous material in the form of choppings:

α. Cyperus kinds:

Cyperus papyrus L. = papyrus reed
(Sum GAL = Akkad. Urbatu)
Cyperus longus L. = "Pfahlrohr" (= elpitu?)
Cyperus rotundus L.
(the type mentioned by Pliny Nat. Hist. XXI, 69)

β. Arundo donax L. = so-called sweetrush
(Sum. GI-BU = Akkad. Mallilu)

γ. Straw, which was excellent material, particularly if we remember that, in Assyrian times at least, the stalks were cut very short in reaping and the straw was then harvested separately (Assurn. Ann. III, 82).

The best known cereals were:

Paniceum miliaceum L = millet =
(Sum. ŠE-BULUG = Akkad. duhnu)
Hordaneum vulgare L = barley =
(Sum. ŠE-BAR = Akkad. uṭṭaṭû)
Triticum dicoccum Schüb = emmer)
(= Sum. ZÍZ-A-AN = Akkad. buṭuṭṭu)
Triticum Monococcum = einkorn.

The method of separating the vegetable matter from the rest of the mastic was described elsewhere (274). Remains of rushes and reeds were found in nearly all samples, while petrified straw remains were occasionally present. In one sample from Tell Asmar chips of willow twigs were discovered, though no contemporary literature has come down to us mentioning their use.

So far as the little data at our disposal permit us to judge, it would seem that the use of fibrous materials in these mastics was abandoned in the Neo-Babylonian period. This rather careless compounding of mastics necessitated a different form of joint construction to that formerly used in order to prevent the mastic from flushing up and creeping out over the face of wall or over road surfaces.

Before I discuss the application of these mastics we must try to discover their method of preparation. I have already pointed out that the chief raw material for these mastics was the very pure "seyali" from the seepages near Hit, which when mixed with sand and lime, is called "ghir" by the natives.

As Table V will show, this bitumen has a considerable water content and on being heated resembles an unstable bituminous emulsion (a suspension of small particles of water in bitumen). The advantages of these minute suspended waterparticles became apparent, when an attempt was made to reproduce these mastics by taking proportions and materials similar to those found by analysis of the samples. The bitumen from Ain el Maraj (melting point about 64° C) was found to mix very well and easily with the mineral aggregate (limestone, lime or loam) in the different proportions found. If these mixtures are heated longer than an hour, all the water evaporates and the material becomes more difficult to work, although at 120° C to 160° C, after a quantity of fibrous material has been added (chopped straw), all the mixtures are in the correct thickly fluid, trowellable condition and it can be worked to a sufficiently compact mass by trowelling and rolling or tamping. As I have already pointed out from a scientific point of view nearly all these ancient mastics contain an excess of bitumen.

To make a mixture of bitumen and mineral aggregate pourable at high temperatures a certain minimum quantity of bitumen has invariably to be used; for modern mastic this is 12 to 16%, as result the mastic has to be manipulated at 180 to 200° C. In the case of ancient mastics this minimum quantity is usually far exceeded. By this means the mastics are less viscous in the heat or, to put it differently, are pourable at much lower temperature. This is undoubtedly of practicable importance in a country as poor in fuel as Mesopotamia and after experience had established this fact an excess of bitumen was probably general practice. The fibrous material was subsequently added with the deliberate intention of obtaining the results discussed above. A proof of this is that we begin to see an increase in the percentage of vegetable matter side by side with an increase of the bitumen (or more correctly an excess) or an increase in the percentage of coarse particles of the mineral aggregate. The percentage in question need not necessarily have been deliberately calculated, but adjustment of the consistency of the heated mass to a satisfactory consistency for pouring and trowelling, etc. would in itself roughly lead to it. Here again a much later text may help to illustrate the method of manufacturing mastic. We refer to a passage from al-Kazwînî (194) which runs thus: "There are two kinds of alkir (native asphalt), first the kind that oozes from the mountains and then the other ḳind that escapes with water from certain pools, it boils together with the water of the spring and as long as it remains in the water it is soft. If we separate it from the water it cools and dries. It is extracted by means of mats and thrown on the shore. Then it is put into a kettle which is heated, the adhering sand is dissolved (mixed) and more sand is added and stirred to a good mix. Afterwards when the mix is ready, it is poured on the floor and becomes solid and hard. Ships and bathrooms are also painted with this mix."

These mixtures were undoubtedly prepared in earthenware jars or pots, not too big for handling. Curiously enough, I found in a Tell Asmar mastic (sample C, Table IV) a potsherd of coarse earthenware, the outside of which was finished with a fine clay slip, whereas the inside was rather porous and saturated with bitumen, a few particles of carbon still sticking to the inside.

This is not improbably a fragment of a pot used for the preparation of mastic and it is equally not impossible, that some of the bitumen-lined jars, found in Mesopotamia, are really the actual melting pots of mastic. Another proof of my identification or ESIR-É-A as mastic may be found in the fact that this material is sold by the GUR while other types of ESIR are sold by weight (mina or shekel) not by volume. This transpires from the examination of contracts from the Ur III and many other periods (275).

In the same sample a well-preserved beetle was found, which with much well-preserved vegetable material (which would have been

TABLE V

Bitumens from surface deposits in the Near East

Origin	Syria			Palestine		Mesopotamia				
	Lattakia	Sükel Chan: near Hasbaya	Cham	Dead Sea		Ain el Maraj	Ain Ma-Moura	Hit	Abu Gir	Bahrein Island
Type	Rock asphalt	Glance Pitch	Glance Pitch	Glance Pitch	Asphaltite	Bi-tumen	Bitumen	Bitumen	Natural asphalt	Rock asphalt
Composition:										
% Bitumen	6.2	ca 100	ca 100	?	98.0	79.0	72.0	64.0	86.5	22.8
% Mineral matter	93.8				2.0			0	10.0	76.6
% Water						21.0	28.0	36.0	3.5	0.6
Analysis bitumen										
Melting point R & B method °C	100.5	Fuses at 135° C	—	Fuses at 135° C	130	64	52.5	47.5	127	
Penetration at 25° C	$1^1/_2$	—	—	—	—	25	73	108	—	
Ash %	1.26	0.5	0.24	—	2.0	3.8	0.7	0.5	—	
Sulphur %	4.27	0.4	9.0	—	6.5	8.8	8.5	8.3	7.3	
Mineral matter	Dolomitic Marl + 30% clay							Gypsum prominent in ash	Ca, Mg, Fe, traces of Van.	Ca Al silicate
Remarks		Brownish black Brilliant lustre	Black Bright lustre, con-choîdal-fracture	Black Brilliant lustre conchoîdal-fracture		Black semi-solid bitumen containing much emulsified water			Lumps of dark brown material	
Reported by	Shell A'dam	Abraham (6)	Abraham (6)	Abraham (6)	Fischer in (127)	Shell A'dam			Shell A'dam	Abraham (6)

charred at temperatures higher than 180° C) goes to prove, that the final temperature of the mixture (120°—160° C) which we established empirically corresponds to actual fact. The voluminous slurry thus obtained must then have been applied to joint, wall or floor and afterwards smoothed, trowelled or tamped in the same way as in modern practice.

According to a statement by Professor Frankfort, several "trowels" of earthenware were found in Mesopotamia, consisting of a round or rectangular disc into which a handle was baked.

TABLE VI

Asphaltic bitumens produced from Near-Eastern crude oils

ORIGIN	MESOPOTAMIA			IRAQ			PERSIA		
	Quayarah	Baba Gurgur							
Analysis									
Melting point R and B Meth. °C	43.5	41.5	50.5	40.5	51.5	87	43.5	50	106.5
Penetration at 25° C, 100 g, 5″	139	195	47	183	45	5	109	45	3
Ash %	0.18	—	—	—	—	—	0.07	0.08	0.05
Sulphur %	7.2	3.0	3.2	5.3	5.9	6.0	2.5	2.7	2.0

That these generally show no traces of bitumen may be due to the fact, that to prevent the mastic from sticking to the implements, these were first dusted with loam or the like. When compared with modern mastics, these ancient mixes are found to be very similar (Table VII).

The standard appears to be approximately that of a mastic with about 25% of bitumen and a small percentage of fibrous material, besides mineral aggregate having about equal portions of a sand and a filler.

This composition varies only slightly during the whole period of Mesopotamian history, as far as we may conclude from the samples analysed so far.

2. *Bitumen as a building material*

After this short discussion of the ancient mastics and their composition I must return to the application of bitumen and first of all discuss its most important use, viz. *as a building* material, a use practically confined to the Mesopotamian civilisation, for bitumen is used only in a few cases in Mohenjo Daro and Köhler's statement that it was used as a mortar in building the Pyramids is incorrect.

The great importance of bitumen (or mastic) for building purposes in these regions was partly caused by the fact that Mesopotamia suffered from a shortage of natural quarry stone and even of timber. Andrae has demonstrated very clearly in his study entitled *Das Gotteshaus und die Urformen des Bauens im alten Orient* how architecture devel-

TABLE VII

Composition of bituminious mastics for different purposes

Purpose	Floor-Mastics				Roof-mastic	Mortar for joints	Mortar or plaster on concrete
Bitumen content %	12	15-17	14-18	15	15	12-16	15
Sand content (coarse fraction) %	37,5°	50+	46-22°	25-30	40+	50°	45°
Filler content (fine fraction) %	51,5	50	40-60	60-55	45	50	40
Reference	(6)	(6)	(6)		(6)	(6)	(6)

\+ Size down to 0.177 mm
° Size down to 0.074 mm

oped from primitive methods using reed bundles, mud bricks and mud plaster, making use of the special properties of quarry stone or wood seldom, and then for integral parts only. It was only in the northern parts (Assyria) that natural stone occurred in any quantity, but even so it was seldom employed for building purposes. The southern part of Mesopotamia, Babylonia or Sumer and Akkad, is an alluvial valley, the inhabitants of which could only obtain stone by barter with the inhabitants of the surrounding mountainous country, with whom, however, they were perpetually at war. The same holds good for timber, for the wood of the datepalm is only a very poor substitute for cedar, oak or fir. It is therefore very logical that in this country natural stone was only used for those purposes where it could not be replaced by the ordinary burnt or sun-dried bricks, e.g. for thresholds, sills and the like. In general practice sundried mud bricks were used for common houses, the use of burnt bricks being confined to the construction of palaces, temples and the like.

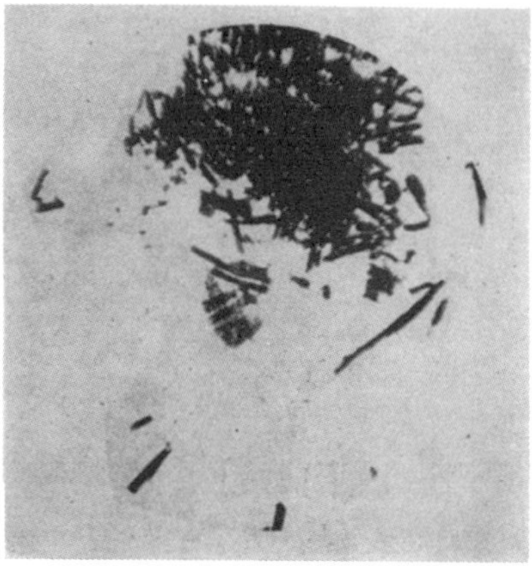

Tissue of a species of Salicaceae (willow) (from Sample H).

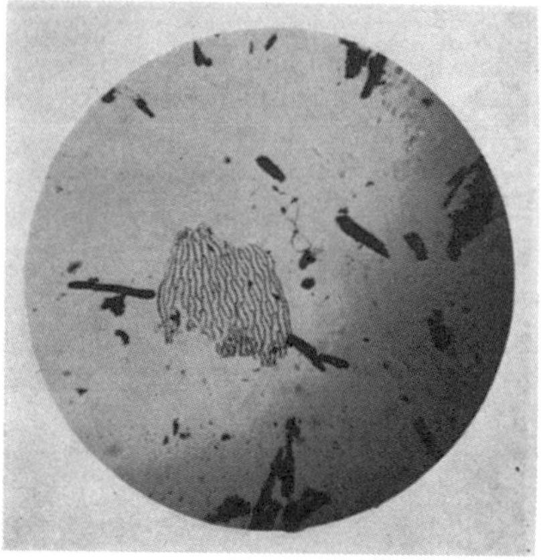

Sample of peat-moss tissue (from Sample K)

Fig. 5.
Vegetable remains in ancient mastics.

Fig. 6.
Repairing the asphalt surface of of New Street, Baghdad.
(Photograph R. W. Parkhurst, 1930).
Mixing the mastic (”ghir“) from ”seyali,“ sand and lime.

The subject of the earliest forms of the bricks and their bonding has been discussed in a very clear study by Delougaz (276). In the earliest period the bricks were moulded by hand, a rough tile-like shape being probably the oldest model, though the socalled "plano-convex brick" with its flat and its rounded side is more easily recognizable as an early model. The form and regularity of these bricks was considerably improved when in the Sargonid period, moulding of the bricks in wooden frames placed on rush mats same the general procedure. The oldest of these moulded bricks are still very large (14″ × 14″ × 2½″) but later on smaller sizes were made and the finish was improved. The general difference between these older bricks and modern bricks is found in the fact that the kilns in which the bricks were baked were very primitive and, of course the very high temperature of our modern kilns could not be obtained in them because of the poor quality of the fuel generally used. The baking temperature seldom rose above 550—600° C, as Rathgen has empirically proved (277). The product of these rather low baking temperatures is a soft brick considerably more porous than our present-day brick. Combined with the bituminous mastic, however, the resultant brickwork was not only very solid, but by pointing the brick with this mortar, the compressive strength of the brick improved substantially as the result of the absorption of a large quantity of mortar by the brick.

We have already referred to the irregular size and shape of the earliest bricks, and this non-uniformity involved, of course, correspondingly irregular and sometimes very large joints. Now these thick layers of mastic between the bricks would tend to flow off or to being pressed out under the weight of the superimposed brickwork, if the mastic were composed of pure bitumen or a compound with sand and loam only. Here arose the early use of fibrous materials in the mastic ensuring a perfect resistance against flow.

As the technique of brick manufacture developed bricks of a more regular form, it was possible to obtain smaller and more regular joints. For instance, whilst the joints between the older "planoconvex" bricks were 3 to 5 cm wide, those in the neo-Babylonian buildings of 600 B.C. are only 1 to 1½ cm.

This is also shown by the particulars on the general appearance of the samples of ancient mastics mentioned in Table IV. The narrow joints of neo-Babylonian brickwork may have been one of the reasons for discontinuing the use of fibrous materials in the mastic. It is possible that Sample L (Table IV) is an example of the bitumen-lime mixture which became more current in that period.

When in this later period the walls were covered with a flat ornamentation of an outside course of beautifully glazed bricks, so popular at that time, a special form of pointing was employed, the joint being only partly filled with the bituminous mortar, in order to preclude the possibility of the bitumen staining the glazed brick. Nevertheless Koldewey—to whose careful excavations we owe the splendid reconstruction of Babylonian Processional Roads in the Pergamom Museum

Fig. 7.
Bitumen mortar of a corbel-vaulted chamber.
Ur, Mausoleum of Bur-Sin, 2000 B.C.
(Br. Museum Photo no. U. 1676).

—found that "it was exceedingly difficult to separate the brick courses from each other".

Towards the end of Nebuchadnezzar's reign, bituminous mortar was abandoned in favour of a much less stable lime mortar to which varying quantities of bitumen were added. After the collapse of the Neo-Babylonian Empire, the new rulers, the Persians and Seleucids, abandoned even this addition of bitumen to lime mortar and confined themselves generally to the use of loam mortar, a decidedly retrograde step. The very firm bond between brick and bituminous mortar was already manifested to the classical writers, indeed many of them mention this type of brickwork. Xenophon (278) and Diodor tell, full of admiration, of the Median Wall, as Herodotus did before them. Vitruvius (279) and Strabo (280) add that this brickwork was also used for vaulting and arches. Cassius Dio tells us the following (281): "There in Babylon, Trajan saw the asphalt with which the walls of Babylon had been built (for, together with bricks or gravel, it produces such strength that the walls made of it are stronger than rock and any kind of iron)."

This was also endorsed by the first archaeologists who worked in Mesopotamia. Among other things. Layard says that "the bricks

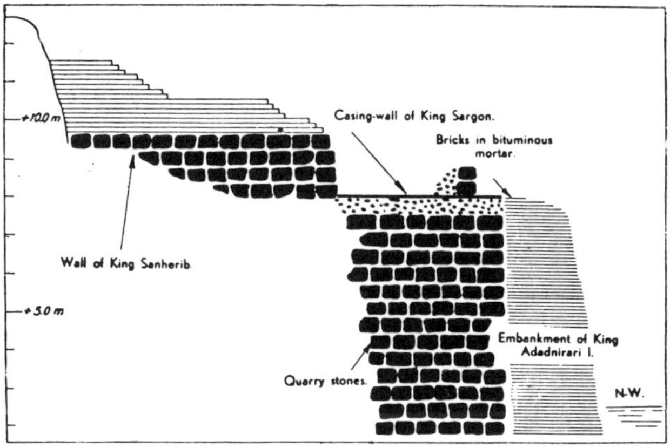

Fig. 8.
Embankment at Assur.

bonded with asphalt have remained immovably in place for thousands of years."

But even to the Babylonians themselves this combination represented something which men referred to as a symbol of stability. This is evident from a current proverb for, instead of saying: "Everything goes against me", the ill-fated person would complain: "Now, here I am in a house built firmly of brick and asphalt, in spite of which a lump of loam falls on my head" (282).

A few examples will convince us that this kind of brickwork was common throughout the entire history of Mesopotamia. The oldest known houses in the region were revealed during excavations at Al'Ubaïd. They consisted of a simple frame of arched bundles of reeds to which rush matting, coated with bitumen, had been attached to form the walls.

The prehistoric temple mound in Erech discovered by Andrae is of a slightly later date. This platform, 12 metres high, was erected to form the foundation for a temple of Anu. It had been built of lumps of pounded clay intersected by courses of dried brick and bitumen for the purpose of strengthening the clay mound and preventing it from drying out. It is but a small step from this primitive construction to the massive brickwork of later ages. The oldest buildings in "plano-convex" bricks were found by Taylor in Eridu and later also in Ur. Bituminous mortar had been used everywhere as was the case in other but more recent towns, such as Kish and Nippur.

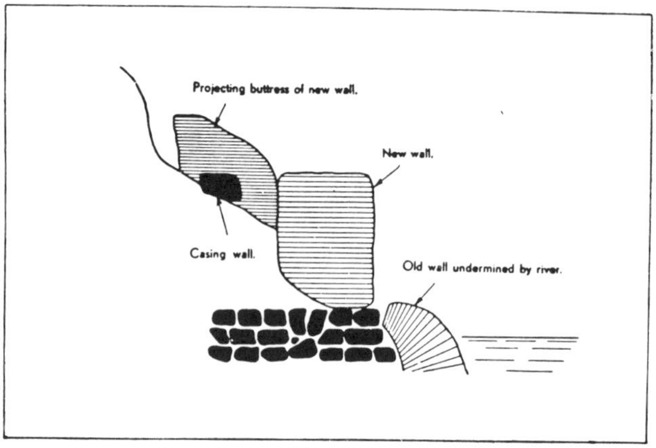

Fig. 9.
Repairs to embankment at Assur.

In Nippur quarry stones bonded by mastic were used for the building of the oldest walls and gates, while the oldest houses are still built with loam mortar.

In Ur the Third Dynasty kings left us many splendid examples of corbel-vaulted chambers, etc. bonded with mastic (283).

In the city of Babylon bituminous mortar did not come to be used until the time of Hammurabi, as Koldewey and Reuther have proved (284).

A "revolution" occurred in Neo-Babylonian times when a considerable change of technique seems to have been caused by the extensive constructions or rebuilding projects of Nabopalazzar and Nebuchadnezzar. As far as we can judge, the use of fibrous material in the mastic was discontinued and a new method adopted. Each course of bricks, after being pointed with mastic, was first covered with a thin layer of the same mastic. To this was then applied a thin layer of loam, which carried the following course of bricks. In this way joints 1—1½ cm wide were obtained. At the same time, every fifth joint was provided with a layer of bands of reed culms beaten and thereafter plaited into mats, which were apparently intended as a species of reinforcement. It is not clear why the architects of this period departed from the original good technique and prevented adjacent courses of bricks to be cemented together by applying a layer

Fig. 10.
The water tank in front of a temple discovered during excavation at Mohenjo Daro (Indus Valley), showing the oldest known use of bitumen as dampcourse, 2300 B.C. (Photo Archaeol. Survey of India).

of loam. The layer of matting, already mentioned by Herodotus (285), has been excellently preserved, wherever it lies embedded in the mastic.

Timber was only sparsely used in architecture. A palm-wood column overlaid with mosaic of red-ribbed stone and mother-of-pearl in mastic was found at Al Obeid. Asphalted trunks of trees were often built into the masonry. Some examples of this were found in the 2 metre thick walls of the temple tower of Ninmach in Babylon. Corners of walls were also strengthened with asphalted poplar-wood.

In vaults of arches, the joints between the lower bricks were made without mortar or with a mastic-loam composition, but the keystone was first dipped in bitumen before being put into place. Older vaults, such as for instance a corbel-vaulted chamber at Ur dating from the Third Dynasty, are executed in the normal mastic brickwork.

3. *Bitumen as a water-proofing agent*

The excellent water-proofing properties of bitumen were put to good account in very early times, and many examples testify to the appreciation by the ancients. Even in prehistoric times *bitumen was used as a water-proofing agent* for example for the brick basins at Ur and at Erech (286).

Fig. 11.
The wall of the water tank at Mohenjo Daro, showing the bituminous layer. (Photo Archaeol. Survey India).

At Al Obeïd drain-pipes were mended and coated with bitumen. In Assyria bitumen-lined bathrooms are already in use during the Tepe Gawra VIII period.

Another very old use of bitumen as a water-proofing layer occurs in the valley of the Indus, where some fourteen years ago a number of relics of an ancient unknown Pre-Aryan civilisation were found. Sir. J. Marshall estimates that this civilisation reached its peak between 2500 and 2000 B.C. and declined before the well-known invasion of the Punjab by the Aryan tribes.

In front of a temple in Mohenjo-Daro, a bath has been found which was very probably used for ritualistic cleansing purposes. This bath measures 39′ × 23′ × 8′ and on the outside of its 3—4 feet thick walls, as also under its floor, a 1-inch thick layer of rock asphalt had been applied between two courses of baked bricks. The three large supply and drainage channels had been water-proofed in a similar way (287). In Nippur in Mesopotamia, at about 2500 B.C., gutters were built of brick set in a bituminous mortar. A little later, about 2300 B.C., the ziggurat of Enlil in Nippur was built, and its sub-

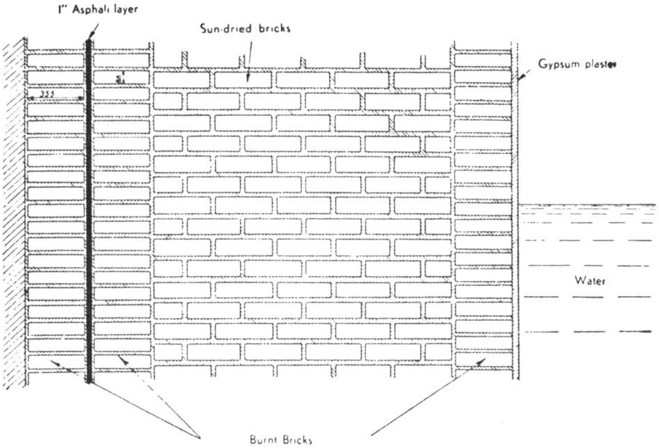

Fig. 12.
Cut through the Wall of the Watertank at Mohenjo Daro (Indus Valley).

structure was insulated by the application of a protective coating of bitumen to the sloping outside wall. In Khafaje gutters were found which were constructed as follows. A groove, 70 cm wide, had first been dug and filled with loam, the desired gutter was then cut into this layer of loam, after which a mastic layer about 3 cm was thick applied to the finished mould (abt. 1900 B.C.). In Assyria, it is a very general thing to find that bituminous mastics were used as damp-courses under floors, and as linings to drains. Bononi mentions such waterproof layers in buildings at Niniveh (288). In Bismayah, too, gutters and drainpipes were coated with mastic or actually made of it (289). The use of such mastic is very general for the construction of wells and waterbasins, bathrooms, in fact everywhere where the bricks could be damaged by rain or by running water. Examples abound at Babylon, Khafaje, Ur (shrine of Bur-Sin), etc.

These useful properties of bitumen were also used extensively in the building of embankments, quay-walls and dykes. Mesopotamia was a country, which could only sustain any considerable number of inhabitants by the careful upkeep of the system of irrigation. The dykes and embankments also served to protect the country from the torrential flood waters arising from the melting of snow in the mountains of the North.

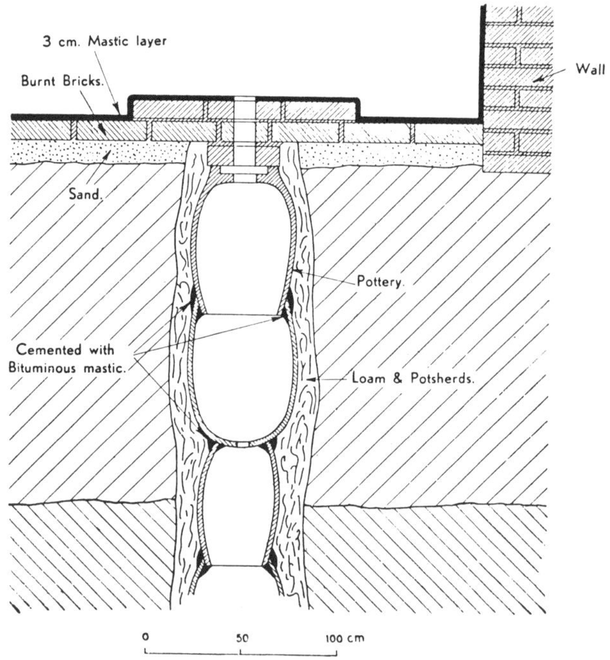

Fig. 13.
Kassite Drain with Mastic Floor.

These embankments and dykes were constructed with very great care and several examples have remained intact upto the present time, including some very fine ones in Assyria. In Assur, Adad Nirari I built in 1280 B.C. an embankment 5000 feet in length along the banks of the river Tigris. In the construction of this embankment, which was of course carried out behind a temporary dam, a retaining wall was first raised, this wall being faced with limestone blocks jointed with mastic, and in its turn, this rough masonry was protected by an outer wall of brick keyed to the main body of the structure by means of buttresses or counterforts at intervals, the buttresses even connected up behind the wall, and built as an integral part of it. These buttresses were 5 feet in thickness and extended backwards for as much as 20 feet. The whole of this brickwork was jointed with bituminous mortar (290).

In this embankment some apertures have been found containing foundation deeds.

We can read in these tablets that the bitumen (the texts say "kupru" = pitch) used was mined at Ubasê (the present Qal'a Shargat). According to Andrae, use was made partly of natural bitumen and partly of a mixture of bitumen and loam or bitumen, sand and gravel. In his description of this embankment Andrae observes "that after 3300 years it still faithfully fulfils its purpose." (291). Meissner informs us that the great Assyrian king Sanherib stated in one of his inscriptions that he was "covering the bed of the diverted river Telbiti with rush (matting) at the bottom and quarried stone on top, bonded with natural pitch. I thus had a stretch of land, 454 ells long and 289 ells wide, raised out of the water and changed it into dry land" (292).

Similar embankments were built in Babylon by Nebuchadnezzar (293).

It is in this Neo-Babylonian period that the number of uses to which bitumen was put in structural works greatly increased. One example of this is in connection with the bridge, 370 feet long, which Nebuchadnezzar built on the Euphrates near Babylon, incidentally a work, which Herodotus erroneously ascribes to the legendary queen Nitrocris (294). The piers of this bridge, 9 metres high and 21 metres wide, were placed at intervals of about 9 metres. They were constructed entirely of brick in bituminous mortar, and moreover the base of each pier was provided with a protective coating of bitumen. As another example may be cited the walls of the Royal Palace in Babylon, the footings of which were protected by a mastic layer on top of a rush mat (295).

Nebuchadnezzar also built large "cloacae" for the drainage of the city of Babylon (296). In one case, he constructed a channel through the thick layer of loam, which had been cast over the relics of the older Sargonid wall, thus greatly improving the drainage from the inner city into the Euphrates. This channel was lined with "blocks of asphalt" consisting of a mixture of bitumen, loam, and gravel, these blocks being made by pouring the bituminous mixture into a mould placed upon a large tile. On removing the mould, the tile remained adhering to the asphalt block, which was then built into the channel in such a way that the adhering tile formed the outer layer.

Furthermore during this last period in Babylon, the use of bitumen for the jointing of drainpipes became common practice, as also for the lining of water closets. Gutters were often let into the walls of the

Fig. 14.
Mastic covered toilet-seat, Tell Asmar (Akkadian Period).
(Photo Orient. Instit.)

houses to drain water off the flat roof to the street or to a sewer and these gutters were often coated with bitumen (297). The water-tight basins or tanks in inner courtyards of the Babylon of this period are often distinguished by the application of a thin layer gypsum plaster over the waterproofing mastic layer. In the long run, this gypsum

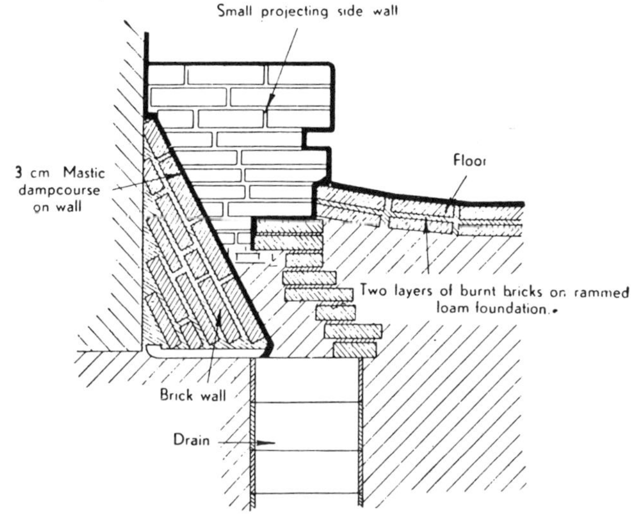

Fig. 15.
Water Closet in Neobabylonian house (showing application of bitumen).

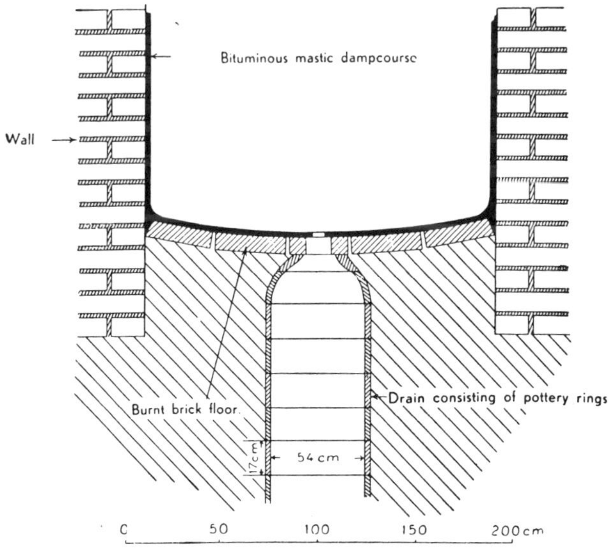

Fig. 16.
Neobabylonian Bathroom.

coating would have been useless for waterproofing but must have mainly been intended to improve the appearance of the work. Andrae found similar examples of somewhat earlier date in Assur (298).

4. *Bitumen as a road-building material*

The system of building with bricks and a bituminous mortar as previously described had been in use for many centuries before it was applied to road building, although it was employed for the construction of terraces for in Nabopalazzar's palace in Babylon we find, superimposed upon ten courses of brick in mastic, a layer of hard

Fig. 17.
Model of Processional Road Aibur-Shabu.
(Photo Vorderasiatisches Museum, Berlin).

core, into the interstices of which mastic was poured, serving as a foundation for the actual floor. Nebuchadnezzar too often built similar terraces for his monumental buildings. He says proudly in his inscriptions: "I made a "nabalu" and laid its foundations against the bosom of the underworld, on the surface of the (ground) water, in bitumen and brick. I raised its roof and connected the terrace to the palace; with bitumen and brick I made it tall like unto wooded mountains" (299).

In a land of caravan roads, the normal needs of traffic would call for no special form of paving and it would be only when the pomp and circumstance of the King and priests increased that special paving was required for the Royal or Processional Road (300). These roads merely connected temples or royal palaces and did not of course form the main traffic arteries in a period without any systematic town-planning. These main roads remained in their primitive state, needing no paved surface, for they carried pedestrians and beasts of burden, for which the pounded or rolled clay soil was quite sufficient.

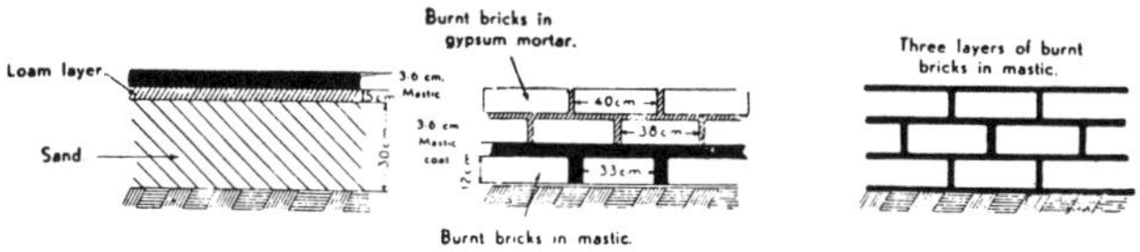

Fig. 18.
Types of Babylonian Floors.

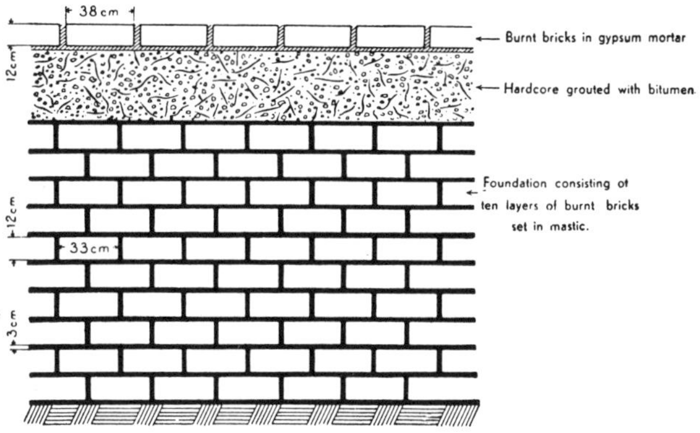

Fig. 19.
Floorconstruction of the Palace of Nabopalassar at Babylon.

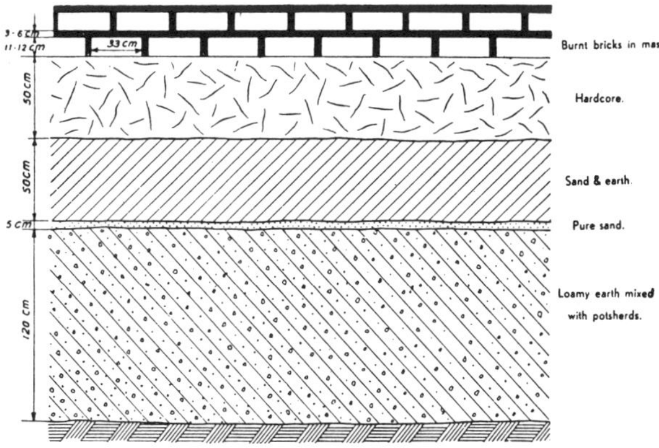

Fig. 20.
Floor of the Temple of Ishtar at Babylon.

The Processional Roads, however, were built with a foundation consisting of several courses of bricks jointed with mastic, and the road surface covering this foundation consisted of slabs of natural stone imported at great expense from the northern mountains. These stone slabs were also jointed with bituminous mortar, and an interesting feature of this work is that the joints were narrow at the top and broad at the base, and only the lower half of the joint was filled with mastic. In this way, precaution was taken against the mastic

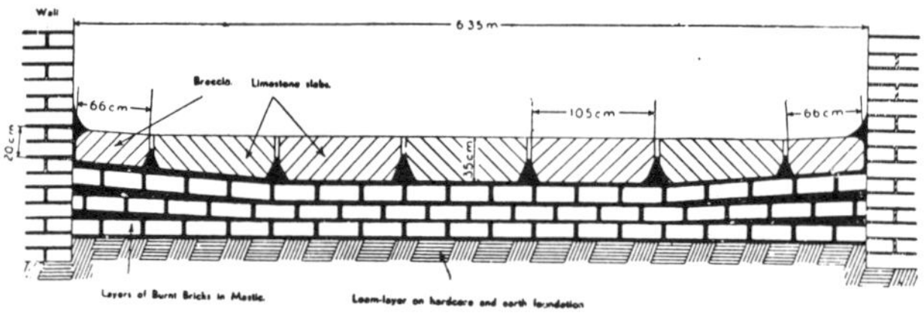

Fig. 21.
Processional Road "Aibur-Shabu" at Babylon.

flushing to the surface, which, as we remember, was especially to be feared because the Neo-Babylonian mastic was no longer stiffened by the addition of fibrous material.

Another feature of these roads is their concave surface, the rain water being drained to the middle and carried away through gullies built into the foundation.

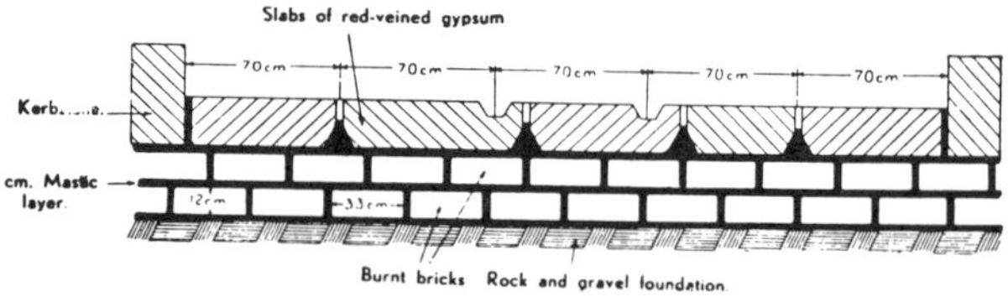

Fig. 22.
Processional Road in Temple of Ishtar at Assur.

The Processional Road in the Temple of Ishtar at Assur shows a surface with artificial wheel-ruts or tracks.

This is a very exceptional feature in Mesopotamia, but it was common in Neolithic Malta and in classical Greece (301). As regards floors, numerous examples have been found in which bituminous mixtures were used. The earliest bituminous floors are simply layers of mastic laid on a rammed loam foundation and examples of this kind dating back to about 1800 B.C. were found in Khafaje, and at even earlier dates in Tell Asmar and Ur. In the city of Babylon, they do not, however, become common until the Kassite period. At a later date, courses of bricks jointed with mastic were used for flooring, and there were very numerous variations employed in this system of laying by combining them with one or more courses of bricks in gypsum mortar.

Our analyses (Table IV) show a number of examples of the use of mastic on floors or threshold. In spite of the hardening of the bitumen through ageing, the cohesion is still fairly good and though several samples are somewhat brittle they still retain a certain thoughness and do not crumble under light pressure. Indeed some of them when tested were still proof to a water pressure of one atmosphere. The earlier samples from Tell Asmar may be called rather thin, when compared with the above mentioned mastic floors of the Kassite period, which are generally 30—60 mm thick. In fact they make the impression of being meant as a water-proofing layer of bituminous plaster rather than a floor. In some cases, the joint between floor and wall is also made with mastic, as examplified in several rooms of the Ninmach temple of Babylon.

These Babylonian roads and floors appear to be very solid from our point of view perhaps too solid, but recently Colberg (302) pointed out, that they might serve as ideal examples of foundations or floors free from vibration troubles.

He states that "the Babylonians have unwittingly constructed these roads to be unimpaired by vibration. The construction of the Processional Road 'Aiburshabu' would serve to protect it against the impact of modern traffic. The above mentioned floor in the temple of Ishtar in Babylon could have served as an ideal floor of modern light steam engine or compressors and the like."

Nissen states that for pavements in Pompeii he found mixtures of bitumen, sand and gravel, but about this modern research has nothing further to say.

5. *Some minor applications of bitumen*

a. Lighting and heating

Both crude oil and bitumen are said to have been used in Antiquity for lighting purposes (303). Although this is not entirely out of the question, the thick sulphurous oil was certainly less suited for illumination than the commonly used olive-oil, and it will also have given less light. No texts give us any information on its use in lamps, but in ancient Babylon, for instance, torches were made by soaking a bundle of rushes in bitumen.

We have already mentioned that in Sicily, in Roman times, 'liquid asphalt' was burned in lamps instead of olive oil and it is not till many centuries later that the lighter fractions or lighter crude oils were

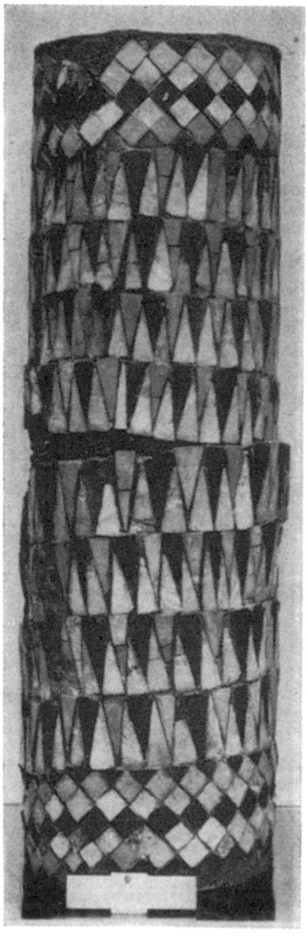

Fig. 23.
Column of palmwood overlaid with mosaic
in red-ribbed stone and mother-of-pearl set in bitumen
(Al Obaîd, 3100 B.C.?)
(Photo Br. Museum).

used. Arabian records tell of naphta torches which the Egyptian Sultan's body-guard carried in about 1000 A.D.

The lack of a suitable lamp for the burning of petroleum, continued to stand in the way of its general use but in a very primitive and unsatisfactory way.

There is little evidence of the use in any early period of petroleum for heating purposes; of course the lack of suitable oil burners would have constituted a serious obstable. Nevertheless, in the later Roman Empire petroleum began to be increasingly used for this purpose, despite the handicap of its inflammability and great volatility. It is therefore all the more remarkable to learn form a description of the city of Constantinople by the curoplastes, Condidos, that 'Median fire' was used as fuel in the time of the Emperor Septimus Severus for the central heating and hot water supply in the thermae (304). The Emperor supplied a great need of the people by building two large thermae, in one of which called "Kaminia", 2000 people were able to refresh themselves daily. Air and water were heated in the subterranean vaults by numerous glass or earthenware lamps filled with burning "Median fire", and by means of these—so say reports—the desired temperature was attained more rapidly than with the usual wood fuel. Unfortunately, these thermae were destroyed during in rebellion a few centuries later.

In the Middle Ages the use of petroleum as fuel again became rare and it is not until the eighteenth that in Baku crude oil and natural gases were once again used for heating purposes.

b. Paints and protective coatings

It was known to classical writers that bituminous paint was used in Babylon. This Strabo informs us (305): "On account of the scarcity of timber their buildings are finished with beams and pillars of palm wood. They wind ropes of twisted reed round the pillars; and then they plaster them and paint them with colours, though they coat the doors with asphalt."

Examples of these pillars of palm-wood found at Erech, Kish and Al Obeïd date back to the prehistoric period and in all these cases the mastic is not only applied as a protective coating, but its adhesive properties are used for decoration of these pillars with mother-of-pearl and limestone inlay-work.

Reuther (306) found that in Babylon outer walls were often plastered over with loam to which a distemper of gypsum was applied, and finally bitumen paint. We are, however, not sure whether the whole of the wall was covered with this paint, or whether it was simply applied as a sort of dado, as the remains of these old walls, which have been found are seldom higher than 5 feet. Bitumen was not only used as a paint in a pure form, but mixtures seem to have been made with chopped straw, reeds, or rushes, on the lines of the normal whitewash of gypsum or lime and chopped straw and similar material. In this way we often find decorations on old Babylonian buildings consisting of alternate strips of bituminous and gypsum whitewash applied on plaster.

Fig. 24.
Engraved shell-plaque with engraved lines and background filled in with bitumen (Ur, Royal Cemetry, 2600—2500 B.C.).
(Photo Br. Museum).

The application of bitumen as a paint actually dates back to the very earliest periods in Mesopotamia. Such paints were not only used for protective water-proofing properties, but also were applied from the earliest times on pottery for decoration purposes. They were for instance extensively used in Susa II ceramics.

Then again, bituminous paint was often used for the purpose of repairing cracked plates, vessels and other articles of pottery, even limestone statues (Ur, III*d* dynasty).

In ancient Mesopotamia it was customary to keep beverages cool by storing them in porous vessels which permitted considerable evaporation, and by this means substantial cooling was obtained. This custom is still adhered to in tropical countries. If such a vessel became too porous or deteriorated in some other way, it was a simple matter to recondition it by applying a coating of bitumen, after which

Fig. 25.
Drainpipe mended and caulked with bitumen
(Al Obaîd, 4th millenium B.C.?).
(Photo Br. Museum).

treatment the vessel, although now unsuitable for its original purpose, could be used for storage and other uses. This method of reconditioning is met with in the earliest Mesopotamian strata. In the earlier periods it was usual to coat only the inside or the outside of the vessel, but in later periods, such as the Persian Empire (from 540 B.C. onwards) both the interior and the exterior of the vessel were coated with bitumen. This was purposely done in the production of cheap pottery so as to avoid the expensive process of glazing, especially in the Seleucid and Parthian periods (307).

A few examples of bitumen used as a protective coat of paint on wooden parts of buildings have also been found here and there in Mohenjo Daro.

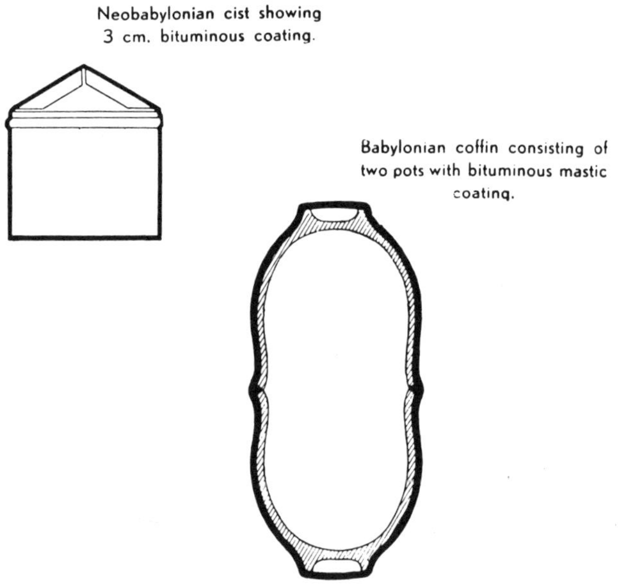

Fig. 26.
Types of ancient sarcophagi.

Bitumen was not used as a paint in Egypt. The so called "bituminous lacquer" found on various boxes, chairs and other objects in Tutanchamun's tomb is very transparant, hence there is certainly no bitumen present and it is as Lucas (308) justly conjectures, a lacquer prepared from a natural dammar resin, which darkens considerably in the course of time.

An instance of the classical use of bitumen as a paint is given by Pliny (309): "The ancients coated the monuments with bitumen, which makes it all the more remarkable to remember that they loved to cover them with gold. I do not know whether it is a Roman invention, but it is said that it was done at Rome first," and he tells us also (310): "It is placed as a coating on copper articles and makes them resistant to fire(?) we have already mentioned that it was once used for staining copper and varnishing statues. Iron workers use it in their workshops for varnishing iron, for the heads of nails, and for many other properties."

These iron-workers perhaps sometimes imitated the Egyptian goldsmiths, for whom we read in Egyptian papyri (311) directions to add bitumen when melting down certain mixtures or alloys. This bitumen is possibly forming an enclosing layer on the molten mixture and may thus prevent oxidation.

One of the most remarkable applications of bitumen as a dyestuff is recorded in the so-called "Leyden and Stockholm papyri" (Papyrus Graecus Holmiensus) which were written in codex form in about 300 A.D., but which actually had their origin many centuries earlier (312). They deal with ancient Egyptian recipes for the colouring of metals, so that on the surface they look like gold or silver, and for the making of imitation precious stones. The basic material for the synthetic gems was a pure silica secretion which occurs in the nodes of the Indian bamboo, known as early as 300 B.C. to Theophrastus and in these documents called "tabasios" or "tabasis" (Arab. tabashir). This secretion is very porous and iridescent and easily absorbs all kinds of substances. This enables it to be treated with various dyes. The procedure was a follows:

The material was first washed in either water, calcium acetate solution (obtained by dissolving lime in vinegar) or rice-water and it was then slowly boiled in a material such as wax. The next process was known as "polishing"; it consisted in boiling in blood, dipping in white of egg and wrapping in a linen cloth for storage.

After this came the actual "staining".

The more usual precious stones imitated were rubies and heliotropes. For the former, mixtures of pitch with resin, "bitumen Iudaicum" or dragon's blood; for the latter, mixtures of wood-tar with alkanna, etc. were employed.

Mumiyâ, the scrapings from mummy linen or the filling of mummies, also went for the making of paint (313) and even in a fairly recent book like that by Newton Friend (314), we find this passage: "Mummy paint is a brown paint used by artists. Genuine mummy paint should consist of dried ground up mummies, mixed with some suitable liquid vehicle such as linseed oil."

c. Water-proofing with bitumen

Bitumen was not only used for water-proofing in architecture or civil engineering for a somewhat lugubrious use of bitumens was for water-proofing and insulating coffins. The chests, in which the dead were laid to rest in a hunched position, were enveloped in a mat and the whole was coated with mastic and covered. The later shapes, also, such as the bath-shaped sarcophagi of baked clay, or the coffins consisting of two large pots, were cemented and insulated with bitumen on the outside. These sarcophagi are usually found in a state of great disrepair, owing to various circumstances. Wherever they have been left undisturbed, however, the layer of bitumen has performed its protective and preservative work most excellently. The mats and the palm-wood lids, protected by the film of bitumen, have then remained wholly unaffected, in spite of the fact that the level of the ground water has risen almost everywhere in Mesopotamia. As Strabo records (315), this great preservative power of bitumen was utilized by the Babylonians, for very often rush or wicker baskets were dipped in hot molten bitumen, by which means they were changed into solid, watertight vessels. A very well preserved specimen of a basket of this kind was found in a loam-lined hole in the foundation of the temple of Ishtar of Akkad in Babylon (316). It proved to have been made of plaited palm leaves and to havc been used as a receptacle for a deed of foundation.

In Babylon bitumen was very generally used for waterproofing the seams and sides of craft. The old saga tells how the great king Sargon was placed as a small baby by his mother, a priestess, in a casket of rushes which she "entrusted to the river, after having closed up the lid with asphalt" (317). This saga became the prototype of similar legends pertaining to important figures in Antiquity; the story of Moses (318) is almost identical with it. But in actual practice the same procedure was followed: a raft or round hamper of rushes or wicker was plaited, which was then waterproofed by stretching animal hide on the outside or coating it with bitumen. To this very day the circular "guffah" which caused Herodotus' astonishment (319) are made in the same way in Hit and Birijik. Herodotus at various stages of his story confuses the "guffah" or "quffa" (Ass. quppu: a plaited basket) and the "kelek" (Ass. kalakku: inflated skin). The quffa is a relation of the prehistoric round coracle, which is still used in Wales and in parts of Inner Asia. It seems that originally the quffa, like the coracle was made watertight with the help of hides or skins.

However in the course of Mesopotamian history more and more "quppu" are coated with mastic. In the days of King Hammurabi (1724—1682 B.C.) the art of making them had become a common trade, for in the famous Code of Law (§ 234—235) we read:

Fig. 27.
Clay figurines of goddesses with bitumen wigs
(Pre-Flood period, Al Obaîd)
(Photo Br. Museum).

"If a boatman caulked a boat of 60 kur for a seignior, he shall give him two shekels of silver as his renumeration.

If a boatman caulked a boat for a seignior and did not do his work well with the result that that boat sprung a leak in that very year, since it has developed a defect, the boatman shall dismantle that boat and strengthen it at his own expense and give the strengthened boat back to the owner of the boat."

A quffa of 60 kur has a volume of 5 register Tons, but the cuneiform texts also mention smaller ones of a volume of 10 or 20 kur and larger ones upto 100 kur (9 Tons). The mastic is heated in a special furnace (kiru), the hot mastic is applied to the surface of the basketwork quffa (tabâku). This part of the work is called "kaparu". A final coat of pure bitumen (ittû) is then painted or poured (nâdu) on the inner side of the quffa. The layer of mastic on the outside is ½" to 1" thick, this agrees with the amounts of mastic mentioned in the texts (about 210 L of mastic per register Ton). The same texts inform us that loads of upto 500 L of mastic are carried by quffa to other cities to be applied to the waterproofing (kaparu) of other quffah. The trade was sufficiently developed to have created its own vocabulary.

The famous Gilgamesh epic (320) tells how the hero, Ut-napishtim, builds a ship in which to escape the approaching flood. He says: "I coated the inside with six Gur (252 L.) of asphalt and the outside with three Gur of asphalt."

This, again, is the forerunner of many similar flood stories. Noah also waterproofed his ship with asphalt (321). Bitumen was also used by the Greeks and Romans for this purpose, although it became increasingly the custom to use tar and pitch in the later Roman Empire.

When reviewing the first edition of this book in JESHO (III. 1960, 218—221) W. F. Leemans stressed the importance of bitumen for caulking and building boats and in view of the important texts he cites (321 a) we reproduce his remarks verbatim:

> Bitumen and mastic found, as is shown by the present texts, a main application in ship-building. Boats were the most general means of transport in southern Mesopotamia and every boat must have been caulked with bitumen and mastic. Pitch was not available there, or at least must have been much more expensive. It may even be said that without the use of bitumen the Sumerians would not have been able to construct their numerous boats, and these boats were a *sine qua non* for achieving the high level which the Sumerian civilization reached. Only by boats of good construction could the metals, stones and other articles which are an inevitable condition for a higher level of civilization be imported in large quantities. The Sumerian trade was based on shipping and boats are mentioned in large numbers and in a great variety of types in the texts.
>
> Caulking boats (*pa/iḫû*) was an important profession. It is surprising that not many economic texts refer to the use of *kupru* and *ittû* in shipbuilding, but this is in accordance with the general phenomenon that only a few economic texts refer to the building of boats. Apparently either texts were as a rule not drawn up in this industry or no texts have been preserved. For some examples from Lagash see A. Salonen, *Wasserfahrzeuge*, pp. 148—149, and A. Falkenstein, *Gerichtsurkunden*, no. 214,57 (esirmá-a). A receipt for 4 kur 2 PI and 3 sutu of *ittû*

(of a certain kind; *ḫizirtum*?) for making 3 boats is recorded in Riftin, *SVIAD* 93, while no. 94 of the same edition records receipts of 30 and 40 kur of the same material (both Old Babylonian).

Some Larsa texts from the reign of Rim-Sin (ca. 1800 B.C.) seem to contain interesting information on the use of bitumen.

1. YBT V 90 (lower part of the obverse of a tablet):

(x lines lost)	
4 (bán) esir *ša a-na li-ib-[bi......] in-na-du-ú*	4 sutu of *ittû* which is laid into the bulk(?) of boats(?),
1 PI I (bán) esir *a-na li-ib-bi* máḫ$^{i.a}$	1 PI I sutu of *ittû* for the bulk(?) of the boats
8 gur esir-è	(totally) 8 kur of *kupru*,
1 PI 5 (bán) esir	1 PI 5 sutu of *ittû*,
ša a-na 2 máḫ$^{i.a}$ 8-gur-ta	which have been put into 2 boats
iš-ša-ak-nu	of 8 kur.

The items of *ittû* are all preserved; these were preceded by items of *kupru*, just as they are in the following texts, to a total of 8 kur.

2. YBT V 231:

12 gur esir-è má da(?)-x-lá	12 kur of *kupru* for ... boat,
4 gur esir-è AL(?).NA.KI	4 kur of *kupru* for...,
1 (bán) esir *a-na ta-al-pi-it-tim ša* ^{gi}-é-má-ra	1 sutu of *ittû* for the cover(?) of the deckhouse,
4 (bán) esir *ša a-na* esir-è *it-ta-ab-ku*	4 sutu of *ittû*, which have been poured on the *kupru*,
1 PI 4 (bán) esir *ša a-na li-ib-bi* máḫ$^{i.a}$ *it-ta-ab-ku*	1 PI 4 sutu of *ittû* which have been poured out in the hull of the boat,
16 gur 3 PI esir-è	(totally 16 kur 3 PI of *kupru*,
2 PI 3 (bán) esir	2 PI 3 sutu of *ittû*,
ša a-na 2 ^{gi}-má 20 gur-ta	which have been put into 2 boats
[iš]-ša-ak-nu	of 20 kur.
Nisannu 8, Rim-Sin 12.	

3. YBT V 234:

15 (gur) 2 PI (bán) ur esir-è	15 kur 2 PI 3 sutu of *kupru*
ša ^{gi}-má 100-gur	for a boat of 100 kur
ša Ṣi-lí-Ištar	Ṣilli-Ištar.
2 PI 3 (bán) *ša* ^{gi}-má-dagal-la $^{i.a}$	2 PI 3 sutu for broad boats
16 gur esir-è	(totally) 16 kur of *kupru*.
2 (bán) esir *a-na ta-al-pi-tim*	2 sutu of *ittû* for the cover(?),
2 PI 2 (bán) *a-na ki-ri-im*	2 PI 2 sutu for the oven (melting pot?),
2 PI 2 (bán) *a-na ni-mir*(?) ^{gi}-má $^{i.a}$	2 PI sutu for the... of the boats,
16 gur esir-è	(totally) 16 kur of *kupru*,
4 PI esir	4 PI of *ittû*,
ša a-na 2 ^{gi}-má $^{i.a}$ 20-gur-ta	with which 2 boats of 20 kur
ig-ga-am-ru	have been filled.
Nisannu 20, Rim-Sin 14.	

4. YBT V 239:

6 (gur) 2 PI esir-è	6 kur 2 PI of *kupru*,
1 PI 3 (bán) esir	1 PI 3 sutu of *ittû*,
ša ^{gi}-má $^{i.a}$ 10-gur-ta	with which boats of 10 kur
ip-pi-ḫi-a	have been caulked.
Nisannu, 25, Rim-Sin 17.	

First it should be examined whether the quantities of *kupru* and *ittû* were loaded in the boats mentioned in the last lines but one of each text, or whether they were used for the construction of these boats. *Šakānu* is a usual verb for loading a boat, but the verbs *gamāru* and *paḫû* are uncommon for loading and the latter verb is found for caulking (CH §§ 234—235). The beginning of text no. 3—a quantity of *kupru* for a boat of 100 kur—seems to show that the *kupru* and the *ittû* were only loaded in the boats of 20 kur, but the exact description of the purposes for which the *ittû* was used in text no. 2 seems not to be relevant for a statement of loading. On the other hand again, it seems unlikely that in the construction of a small boat of 8 kur a quantity of 4 kur (ca. 480 litres) should have been needed (text 1) and for a boat of 20 kur 8 kur of *kupru* (texts nos 2 and 3). In text no. 4 no number of boats is mentioned, but in view of the quantities in the other texts it may be supposed that it was only 1 boat, notwithstanding the plural. It seems that somewhat more than 15 kur was needed for a boat of 100 kur according to text no. 3. From lack of knowledge about the quantities of pitch and similar materials used in the building of similar boats in southern Iraq and on the shores of the Persian Gulf in the present time, no definite conclusion about this point seems to be justified.

Nevertheless the texts yield interesting information:

a. In boat building large quantities of *kupru* were used and only small quantities ot *ittû*;

b. *Ittû* may have been the more valuable material; the quantities are much smaller and these quantities and the purposes for which they were used are exactly stipulated;

c. We learn a number of verbs used in connection with *kupru* and *ittû*: *šakānu*, *gamāru* and *paḫû*. *Paḫû* means in this connection "to caulk", *gamāru* is found in connection with the completion of boats. Other terms applied to *ittû* were *nadû* (text no. 1), *tabāku* and *lapātu* in the II-form (2 and 3).

d. *Ittû* was used in a fluid condition for some operations; it was poured out (*tabāku*) (text 2). In order to obtain the fluid condition it was put into an oven (*kiru*, text 3). It could be poured out on *kupru* (text 2), apparently in order to improve the *kupru*. One wonders whether the purposes in texts nos 2 and 3 were not almost the same and, therefore, whether not only the first but also the other purposes mentioned are perhaps identical.

e. *Ittû* was applied 1) for the deckhouse, 2) for adding on the *kupru* and 3) on the inside of the boat's hull.

f. *Kupru* was probably the more consistent material, but nevertheless it could be measured by measures of capacity, just as *ittû* was. Points d and f are in agreement with the observation of Forbes (p. 21) that *ittû* was the soft, sometimes moist, product of the bitumen-pools and that *kupru* was as a rule harder bitumen or mastic.

g. It may be supposed from the preceding remarks that *kupru* was used for the rougher coating (*kapāru*) or caulking of the boat and that *ittû* was generally used for the finishing touches.

h. The two articles were closely related but yet were clearly distinguished in the period of the texts, and served different purposes.

These important considerations lead Leemans to a discussion of the nomenclature reviewed on our pages 13—22, (including our Table I) on which he says this:

Forbes supposes (p. 18) that ESIR.UD is purified bitumen. In the texts of the

Ur III period this bitumen is found in very large quantities. Another variety, mentioned in large quantities in the same period is esir.é.a, while esir. è.a is also found. All of these were apparently used in large quantities both in building houses and boats. The present texts show that *kupru* was the Akkadian name for the article used in large quantities in boat building. With our present knowledge there seems to be no ground for differentiating ESIR.UD, esir.è (written ESIR.UD.DU), esir. è.a (written ESIR.UD.DU.A). Esir-é-a may have been a quality especially prepared for use in building houses, just as esir-má-a and esir-apin may have been special qualities used for boats and for irrigation works.

In conclusion, for Mesopotamia only the following table seems to be justified (cf. table I):

Genus	species	member	description	Sumerian	Akkadian
Bitumen	a) petroleum	all crude oils	fluid	ì-esir	*šaman-ittî* *šaman šadî* *napṭu*
	b) native asphalts	bitumen	soft and pure or fairly pure	esir	*ittû*
		mastic	impure, large percentage of associated matters	esir-è esir-è-a esir-UD	*kupru*

The quotations under IV (pyrogeneous residues) in the table are questionable.

d. Bitumen as a cement or adhesive

Almost everywhere in Antiquity bitumen was used as a cement. The inhabitants of the lake-dwellings in the Alpine region used it for the cementing of articles of use, for instance to fix arrowheads to the shaft. They also stuck thin leaves, cut out of birch bark, with it to their pottery. Sometimes they would let a groove into the surface of these jars, fill it with bitumen into which they stuck small stones, thus creating a kind of mosaic pattern (322).

The use of bitumen as a cement for various costly gems in mosaics or inlaid work is one which has always attracted attention and which is mentioned and illustrated in most manuals on bitumen. Friezes, consisting of white stone, mother-of-pearl, lapis lazuli or red stone laid in bitumen, sometimes placed in a copper frame, were made as early as the al Ubaid period (i.e., before 2800 B.C.). This technique enjoyed long popularity, and numerous examples have come down to us from the Third Milennium (323). Bitumen is also used for holding gold or other beads (324), fixing knives in their hafts (325), and water-proofing porous material (326).

In prehistoric times ostrich eggs were made into vases decorated with lapis lazuli and mother-of-pearl in bitumen. In this way, shells are fixed to a libation vase of Gudea, and several examples were excavated at Bismaya (327). It used to be the general custom to fix-mother-of-pearl, ivory or coloured stone plaques into the eyesockets of statues by means of bitumen (e.g. finds at Adab, Kish and Eshnunna). In Khafaje bitumen had also been used to fix a heavy albaster relief to the wall, although it was already held by a peg through the hole in the centre.

In Mohenjo Daro bitumen served as a cement to fix wooden boards to the steps of brick stairs leading to the Great Bath. In Egypt it is often used to cement mummy chests, wooden coffins or statues (328). In a prehistoric granary of the Fayum Miss Caton Thompson found a wooden sickle with flint teeth set in bitumen. In Mesopotamia the sickle often consists of a rod or strip of mastic with a row of flint teeth (Babylon, Khafaje).

Another interesting use was discovered in Eshnunna where according to Prof. Frankfort: "We found in the woman's private room of the Akkadian palace the raw material for the Akkadian equivalent of Victorian beadwork and embroidery, sheets of bitumen (mastic?) and shaped pieces of mother-of-pearl with which to inlay them, together with some fragments of the finished work, perhaps intended for the lid of an ointment jar" (329).

A remarkable but much later use is recorded by Lucian (330). According to him the Roman wax-tablets were sealed with "kollyrion", a mixture of Bruttian pitch, bitumen, quartz powder and natural mastic.

There was another application of bitumen which was far more common than is generally thought. The art was apparently understood of beating thin metal bands round sculptured cores of bitumen or mastic. The metal foil was made by hammering the metal out on flat stones and was then moulded round beads of wood or bitumen. This is how they made golden horns of the bulls' heads found in Ur, to which we may assign the date of 2500 B.C. (331). The same technique is evident in the copper bulls of al Ubaid (332). Copper foil was hammered on to a wooden skeleton coated with bitumen. Sometimes, too, a cast-metal object was filled with a core of mastic. Such work is seen in the early cast-copper objects of Lagash and Tello (333), and the same technique was applied in the north (Nineveh) (334). It was even done in Egypt apparently; at least GSELL (335) mentions a bronze

Egyptian statue (XIX or XX Dynasty, ca. 1000 B.C.) in the British Museum still containing a core of bitumen and sand reinforced by a piece of iron.

Bitumen, in the form of natural asphalt or mastic, played an important part in applied art. These two kinds of bitumen were easy to handle, and the results obtained by the Sumerian artists with this material are to be seen in the Louvre, for instance. It is a striking fact

Fig. 28.
Small bitumen basin, Neo-Sumerian period, Susa, second half of third mill. (Louvre Cat. No. 224 bis). Ibex as symbol of fecundity and fertility.

that the sculptural work done by the Sumerians in natural stone in the Third Millennium gives little hint of the artistic talent displayed in their plastic work with easily managed materials like the bitumen compositions mentioned or the wax core of an object cast in metal. This is quite comprehensible for, natural stone in Mesopotamia being very costly on account of its having to come from the mountains far away, they could have had but little opportunity of practising their art.

Prehistoric vases were discovered at Susa cut put of this rock asphalt into a handsome cone shape (Susa I). Another object of great antiquity is a votive relief of Entemena, the patesi of Shirpulak. This tablet has been cast in a mould consisting of a synthetic mixture of bitumen and loam which has now become as hard as a stone. Similar hard mixtures are made even to the present day notably in the shape of a ball subsequently fixed to a stick for use as a bludgeon. De Sarzec found a similar ball in Kish, attached to a copper tube; probably it was used as a flagstaff.

An asphalt ring of the same date (about 2450 B.C.) was found in Ur, it proved on analysis to consist of an artificial mixture of bitumen, sand and filler (336).

Bitumen was used for the shaping of wigs on prehistoric figurines at Tell Obeïd; a similar statue of Parthian date was found at Babylon.

Small beads and vases made of rock asphalt were also found in the lake dwellings of Switzerland. In all likelihood Travers asphalt was used in these cases. In Roman-British days, and may be even earlier rings, armlets, ornaments and parts of furniture were made from the Kimmeridge shale of Dorset (336), but their use was local and did not form part of British exports.

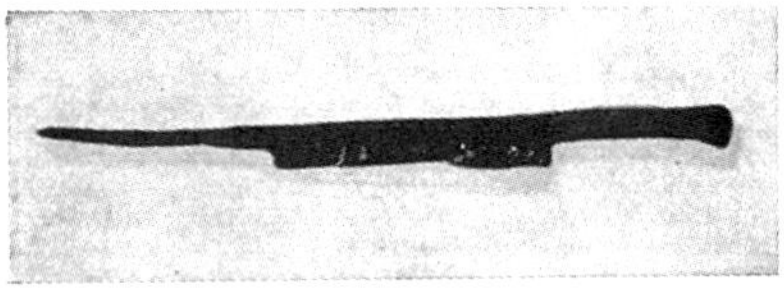

Fig. 29.
Prehistoric Egyptian sickle found in a granary in the Fayum (3000 B.C.)
Flint teeth set in bitumen, wooden handle.
(Photo Caton Thompson).

6. *Bitumens in magic and medicine*

a. Magic

From the earliest times bitumen seems to have attracted the attention of those people who sought to remedy the physical and mental ills of the human race. Often there was no very clear distinction between the medicinal and the supernatural and it is mostly impossible to separate them here as physical ills were often attributed to spiritual causes.

Bitumen plays a double, rather contradictory, part in magic. On the one hand it is used to keep out evil spirits (kupru) (337); on the other, the black substance is regarded as the power of evil (338), and

ESIR is the symbol for, or actually is, the substance in which the demon Ašakku chiefly elects to reside. Bitumen, holds an important place in many magic formulae or rites. Images of the persons or objects to be bewitched were cut out of bitumen and, after the proper incantations, were buried near the victims. In a recently published collection of incantations (339) there is mention of casting off a spell caused by an image made of bitumen (iddû) (ii, 148—159) and casting off a spell

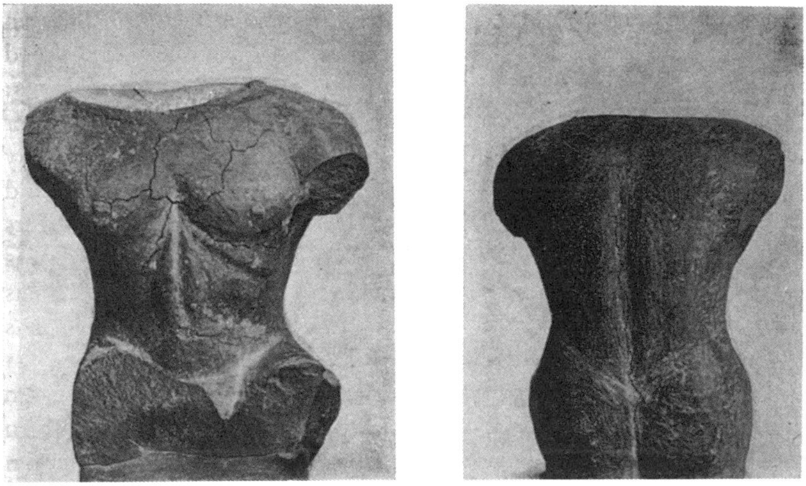

Fig. 30.
Torso from Susa, bituminous stone, third mill. B.C.
(Photo Louvre, Cat. no. 215).

caused by means of a plastered bitumen image (iddû), (ii, 181—205). In another text (340) instructions are given to bury an image made of bitumen under the victim's front door after pronouncing the proper charm.

Another rite demands that the effigy of the victim shall be sacrified to the fire demon (so, shall be burned). The materials to be used are enumerated with great precision, bitumen once again appearing frequently. This form of sorcery was very general in classical antiquity. For instance, in love enchantments the hay or straw torch or the laurel branch used for the spell is ignited with asphalt.

It was a very common habit in ancient days to coat doorposts with bitumen or wood-tar pitch during certain festivals, particularly the Anthesteria, with the object of keeping out spirits.

Bitumen recurs again and again in the magic formulae themselves, and citation of its special qualities is intended to make the formula recognizable to the invoked demons. For instance, a witch calls up the spirits as follows: Like asphalt and pitch which come out of the depth (341); or casts a spell on the victim in the following words: As bitumen holds a ship, so I hold you and will not let you go (342). To illustrate the evil of witches, it is stated in another text that their spells render the victims as powerless as if their mouth and ears were stopped with bitumen (343).

The phenomena observed in the neighbourhood of oil or bitumen deposits were put to good use for soothsaying. Examples were quoted when we discussed the term napṭu. One text (344) runs as follows: "If there are agglomerations like petroleum or sesame oil borne hither and thither on the surface of the water, mušgarru or disease will lay hold of the land." Another says: "If a pit opens in fallow land and burning bitumen appears, the land will be destroyed." (345). Sorcerers also augured by the shapes formed when bitumen or crude oil was poured out into a basin or beaker of water. One text dealing with this form of augury (346) says: "If when I pour naphtha on water it has the appearance of asphalt, that means fate, the sick man will die."

In Babylonian times mystical characters were painted in gypsum and bitumen on the doorpost of a sick person's room (347). Bitumen was supposed to afford protection from the terrible female demon Labartu, who was believed to drink the blood of children and animals. To ward her off, ointments were prepared from bitumen, earth, lard, fish and certain plants and these were applied by rubbing into the skin. The recital of certain incantations was supplemented by the burning of a bitumen image of the person involved, a practice, surviving up to the present day (348).

Again Aethicus Istricus says in his *Cosmography* (800 A.D.) that armour can be made impenetrable by the method practised by the Amazones, of immersing it in a mixture of bitumen and the blood of a child.

Perhaps the connection always made by medieval writers between pitch and bitumen and the devils and other dark creatures is a survival of a much older tradition contrasting black and white magic and assigning to each its specific ingredients.

Finally, we should not forget the part still played in many of the Apollo shrines of classical times by burning natural gases and other manifestations observed wherever bitumen is found (349).

A curious "ethical" use of bitumen was prescribed by the Assyrian law, which decrees, with the cruel characteristics of this people, that for certain transgressions hot bitumen is poured over the head of the delinquent (350).

b. Medicine

Bitumen also holds a prominent place in medicine. We have a few prescriptions (351) in which crude oil (IÀ-KUR-RA) appears, and it is even used for anointing (352). But there are far more prescriptions for the use of bitumen in solid forms. Thus, mixed with sulphur and other substances, it is recommended for fumigation (353) and for the treatment of sores on hands or feet (354). Bitumen (iddû) fluxed with oil is prescribed for inflamed eyes (355). A mixture of beer and bitumen is suggested as a cure for some ailments (356). In describing the symptoms of a certain disease it is stated that the body becomes as black as bitumen (357). It is stated in one curious text that certain diseases can be cured by making the patient stand on bitumen (kupru) (358).

The Assyrians used bitumen externally on a festering foot or finger. The "snakes of asphalt" recovered from the river were used with tragacanth on a pledget of red and white wool in menorrhagia.

In classical antiquity the bitumen found near the Ararat, or, according to some that of the Djebel Djudi (Djudi Dagh) to the southwest of Lake Van, was credited with special healing power. Legend had it that the Ark had landed there, and that the bitumen collected from the rocks there had even greater cleansing power than that which had always been attributed to bitumen from all sources (359). It was also made into amulets.

Pliny's text will serve to give an impression of the different uses to which bitumen was put in pharmaceutical prescriptions (36): "It checks bleeding, has resolving power and assists the drawing together and healing of wounds. The smell of burning bitumen drives snakes away. The Babylonian variety is said to be efficacious in treating cataract and leucoma, as well as for leprous spots, lichens and prurigo. It is applied also as a liniment for gout. All kinds, however, are useful for straightening out eyelashes which inconvenience the eyes. Bitumen, rubbed with soda, cures aching teeth.

Taken in wine it soothes a chronic cough, and relieves shortness of breath. If administered in the same way as to those suffering from dysentery, it checks diarrhoea. A potion of bitumen and vinegar dissipates and removes accumulations of clotted blood, and gives relief in cases of lumbago and rheumatism. A poultice with barley flour has special merits and is named after the bitumen which it contains. It checks bleeding, promotes the closing wounds, and draws served muscles together. For quartan fevers, the prescription is a drachm of bitumen and a like weight of mint, triturated with an obol of myrh. Burning bitumen leads to the detection of epileptics. The smell of a mixture of bitumen, wine, and castoreum dispels congestion of the womb, and fumigation with bitumen checks prolapse. Taken in wine it hastens menstruation."

Bitumen is likewise praised as a panacea by Dioscorides (361), Josephus (362), Celsus and Galen; the latter specially referring to the properties of "gagates" in this respect. Varro (363) describes the disinfecting properties of bituminous materials and their vapours, which he employs to fight various diseases, caused, as he avers, by invisible tiny creatures. Virgil recommends its outward use against scabies (364); but the Geoponica (365) makes sufferers from diarrhoea swallow it.

Moreover, then as now, opinion in the profession differed as to the efficacy of some medicines, and bitumen was an example of this. Thus Philostratus (366) warns against water contaminated with bitumen, as otherwise the intestines will stick and close up. Nevertheless, traces of bitumen appear to have been found in Roman wine and water bottles, although not in Egypt as Neuburger claims. Perhaps it was used here as a preservative or to improve the taste, as Pliny claims for resin (367).

Again Aetius of Amida is of opinion that the drinking of bituminous water is the best medicine against dropsy (368).

For use in medicine great claims were made upon the purity of the bitumen, and colour, weight, etc. were carefully considered. As a curiosity we shall here quote a prescription taken from the *Tetrabiblos* of Aetius of Amida (500 A.D.), according to which a plaster could be made against tumours, swelling and gout as follows: "Take 360 drachms each of terebinth oil, "Stone of Asia", and of bitumen; then 120 drachms of calcined soda, calf's fat, wax, bay leaves, ammonia and thyme, pyrites and slaked lime; to this add the ashes of snakes burned alive, about 140 drachms and 2 pounds of old oil.

First heat the oil with the wax, dissolve the bitumen (previously triturated) in this, after which add the remaining ingredients."

c. Agriculture

Bitumen was burnt (with sulphur) for fumigation or as incense in

Assyrian and later temples. It was also used for fumigation in *agriculture*, where it was burned under trees or bushes to kill caterpillars or other harmful insects.

It is also recommended for disinfecting nests and cages (369). The same collection of recipes for agricultural use contains more prescriptions involving the use of bitumen. A mixture of bitumen and oil is rubbed on to the wounds of trees or it is used to form excellent rings of agglutinant round trees to keep the ants away (370). The bitumen is mixed with the oil by trituration in a mortar or by dissolving it.

Mixed with various spices, samples are prepared from bitumen for cattle plague (371), and if fowls are rubbed with a mixture of bitumen, resin and sulphur, they lay bigger eggs (372).

d. Mummification

It is very generally supposed that bitumen was used in Egypt for *mummification* and *embalming* of the dead. This, however, is only partly true and appears to be due to an incorrect generalised interpretation of passages of Diodor and Strabo by such modern writers as Jeep, Reutter (373), Tschirsch (374) and Wallis Budge (375).

Balsams, resins etc. were generally used for mummification. Extensive research by Lucas has shown that wood-tar pitch has been used since Ptolemaic times (376) and in later work he has not been able to cite the use of bitumen; on the contrary, he disproved Reutter's analyses which identified bitumen. Spielmann tried to detect bitumen spectrographically, but came to the conclusion that "bitumen and resin are present in relatively low percentages, the Egyptians undoubtedly having used wood-tar pitch as a diluent". One has perhaps been tempted to think that bitumen was used because, since the Twelfth Dynasty, the mummies have been dark, sometimes black. Without proof to the contrary, it must for the time being be regarded as certain that bitumen was not used to any appreciable extent before the Ptolemaic period, and even then only for second-rate processes of mummification. We share Coremans' opinion that "it is generally admitted that bitumens, or like products, were "frequently" used for ancient embalming processes. This does not apply to Egypt, where resins, almost to the exclusion of anything else, were used for the purpose. It was not until the Ptolemaic period that bituminous materials were perhaps "very rarely" used. Only a very marked similarity in the appearance of these two groups of substances could have given rise to this regrettable confusion" (377).

One of the causes was the Egyptian medieval trade in "mûmiya", a word which in the Persian language denotes "wax" and in Arabic "bitumen". This word came to be applied to the scrapings obtained from mummy linen or extracted from the cavities of the embalmed bodies and sold as a drug.

A passage from the works of Ibn al-Beitar describes the situation very clearly: "Mumiya is the name given to the drug just mentioned and to the bitumen of Iudaea and to mumia of the tombs as found in great quantities in Egypt and which is nothing else but the mixture formerly used for embalming the dead, in order that their dead bodies might remain in the state in which they were buried and neither decay nor change." This commercial product of widely varying origin and composition came to the West in the 12th century being used in paints and medical prescriptions, a trade which became virtually extinct in the 17th century after having done incalculable damage to valuable mummies, and graves, which might have been spared for a more scientific and careful excavation (378).

Thus a word wrongly interpretated to point to the substance in the dead body gave rise to confusion, the original Egyptian expression for the preserved body "qasiu" or "qas" reveals nothing as regards the use of bitumen. The only classical writers referring to the use of bitumen. The only classical writers referring to the use of bitumen in mummification are Diodor (379) and Strabo (380). Modern research on these substances by Lucas (381) and Spielmann (382) leads to the conclusion that bitumen was probably not used generally before the Ptolemaic dynasty, though the latter favours its introduction since the XXIInd dynasty.

As other evidence shows that the bitumen fishery of the Dead Sea turned out to be of great importance to Egypt, I cannot support Lucas' conclusion that bitumen was not widely used in the Graeco-Roman age. It may have been much more widely used especially in the rougher and cheaper forms of embalming, much evidence being destroyed later on by the Arabs. It is therefore highly desirable to analyse more samples and thus obtain more statistical evidence as to the frequency of the use of bitumen, either in a pure form or mixed with spices, natron, cedar oil and other materials and especially to find out whether it was perhaps connected with the cheaper forms of burial, thus perhaps given support to the statements of Strabo and Diodor.

We have discussed the different Egyptian terms said to denote bitumen earlier in this essay, others may still be buried in the Coffin

Texts or other funerary documents. However, one of the Rhind papyri (383) contains an interesting passage stating that: "Anubis... fills the interior of the skull with mrḥe Ḥr, incense, myrrh, cedar oil and calves fat."

Möller translates this "mrḥe Ḥr" by "Syrian asphalt". This seems to be an equivalent for the Coptic "amrehe" which corresponds with the Greek asphaltos. Again the Demotic "mrḥe Ḥr" corresponds in this bilingual papyrus with the Hieratic "mnnn" which is stated to have come from Phoenicia, Coptos and the elusive Punt. The first two places would in fact eminently suit our statement that there is a good deal of truth in the classical allusions to bitumen exported to Egypt from the Dead Sea. This and further matters on Egyptian nomenclature of bituminous materials must, however, be left to more competent authorities to decide.

7. *Petroleum and Greek fire in warfare*

Human intelligence, even in the most primitive eras, has always been exceedingly fruitful in the invention and improvement of means of attack and defense.

It is, therefore, not unduly surprising that petroleum and similar materials, though usually innocuous and even beneficial substances, should have been, at an early stage in history, adapted for warfare.

Probably the earliest and simplest use of bituminous materials for military offensive purposes is mentioned in the works of Thucydides, where accounts are to be found of the siege of Plataea (384). Fascines were piled up against the walls of the city, and ignited by a burning mixture of sulphur and pitch.

Similar tactics, according to Thucydides (385), were also adopted at the siege of Delium, a notable addition in this case, however, being the use of an airbellows with which to aggravate the conflagration. Other primitive methods of utilisation for military purposes of the properties possessed by products of petroleum are described by Pliny (386), who records that the attack of Lucullus on the city of Samosata (Tigranocerta?) was repelled with the assistance of burning "maltha" and Philostratus (387), who mentions "an oil that once set afire can not be extinguished, and which the Indian King uses to burn walls and capture cities."

In late Roman military circles, a popular bituminous weapon of a rather higher order of invention appears to have been that of an arrow of fire filled with sulphur, resin, asphalt and pitch, and swathed in oakum soaked in crude oil (oleum incendiarum). Books on tactics like Vegetius (A.D. 400) are full of references of this kind.

The next step in the martial development of petroleum is the invention of what became known as "Greek fire", the use of which in varying forms extended of many hundreds of years. According to von Lippmann, it has been noticed that if a mixture of petroleum and finely devided quicklime was exposed to moisture, the heat generated by the contact of the lime and such moisture was sufficient to ignite the mixture spontaneously, particularly if it contained light (volatile) oil fractions.

This principle of spontaneous ignition caused by the addition of moisture was, according to Theophanes, brought to its highest practical application for military purposes by a Greek architect named Kallinikos, residing in Byzantium in about 650 A.D. as a fugitive from the Arabs of the Syrian town of Heliopolis, and who is claimed by Theophanes to be real inventor of "Greek fire" (388).

But the principle must have been known for many hundreds of years. For Livy tells us (389): "The Matrons dressed as Bacchae... rushed down to the Tiber with burning torches, plunged them into the water and drew them out again, the flame undiminished, as they were made of sulphur and pitch mixed with lime."

And in 200 A.D., Athenaeus writes of a magician or sorcerer known as Xenophon, part of whose stock-in-trade was a self-igniting fire ("Pyr automaton"), and in his *Kestoi* Africanus (300 A.D.) mentions a mixture of natural sulphur, resin, naphtha, salt and quicklime, which, by careful mixing and addition of the lime last, and enclosing the whole in a bronze vessel to exclude humidity, air and light, would ignite spontaneously by the contact with water or dew. The noteworthy ingredient of this mixture is the salt, which gave a yellow colour to the flame and which was therefore thought to make it hotter.

Ammian speaks of arrows coated with a similar mixture, which ignited and then discharged with moderate velocity, flare up violently on coming in contact with water. Procopius mentions the use of a like "elaion medikon" in the Italian wars of Justinian. The Greek-architect Kallinikos then is said to have provided Byzantium with the correct formula for the preparation of a self-igniting mixture of petroleum fractions and quicklime, and mixtures of this sort made in accordance with his recipe were used succesfully on a large scale in the Byzantine Fleet. The instrument of projection employed appears to have been Ktebisios' double acting piston pump, invented as long before as 200 B.C. in Alexandria, which instrument was mounted on the prow of the warships. From this pump the self-igniting or, sometimes, ignited mixture was fired at the enemy. The amalgamation of new and old, however, enabled the fleet of Constantine Pogonatus to inflict a smashing defeat upon the Arabian fleet at Kyzikos (A.D. 678). In 717 "Greek fire" was instrumental in the hands of Leo III (717—741) again in defeating the Arabians, who were compelled by its use to raise the siege of Byzantium. Leo III's hand-book on the tactics of warfare contains many explanatory and eulogistic references to the use of "Greek fire", a word of caution being introduced at the same time against the danger of accidental explosion and the consequent necessity for careful handling of the mixture. Both self-ignition or deliberate ignition at the mouth of the syphon are mentioned in connection with "Greek Fire".

It was not long before "Greek Fire" was adapted for use in hand-grenades, the self-igniting mixture being hurdled in stone or iron jars at the enemy by the Byzantine soldiers or squirted at him by means of hand-syringes ("mikroi syphones").

The Arabs profited by their unpleasant experiences of "Greek Fire" to appreciate its efficacy, and subsequently to learns its composition themselves. The earliest mention of its use by an Arabian fleet is the naval battle of Rachid at which the Sultan's well equiped fleet of 25 ships was able to defeat the much larger fleet of the Egyptian caliph (915 A.D.). Though al-Dschahiz claims that the Arabs had used "Greek Fire" since 600 A.D., more trustworthy authorities like Tabari tells us that the Arabian trading vessels in the Indian Ocean had men familiar with the use of "Greek Fire" on board to protect themselves against pirates.

By the Arabians the use of "Greek Fire" is extended to the Chinese, who employed it effectively since the eighth century on their merchant ships against the marauding Malay and Arabian pirates.

The Greeks themselves constantly profited of the weapon which bore their name. When Igor's Russian Fleet, consisting of no less than 1000 small ships, threatened Byzantium many of them were burned, and the remainder driven back by a Byzantine squadron of a mere 15 vessels. A frank and ingenious admission of defeat was made by the Russian fleet on its return home (A.D. 941), the report of the commander stating that "the Greeks have a fire ressembling the lightning of heaven and when they threw it at us, they burned us; for this reason we could not overcome them". "Greek Fire", often called "pir thalassion" or "oleum incendiarium", played such an important part in Byzantine politics, that Gibbon has very aptly described it as the "Palladium of the Byzantine State". The Emperor Constantine VII Porphyrogenitus (922—959 A.D.) devotes a special chapter to it in his manual on statecraft. He claims divine origin for the receipt and urges that its composition should be kept a secret known only to the emperor and a few confidants. In the same book he gives evasive answers for those who might be pressed to reveal the secret.

This policy was indeed successful for a long time and the secret was not known in the West until very much later documents such as the *Mappae Claviculae* (820 A.D.) mention only fire rockets or arrows from Byzantine sources, but cannot give the formula for "Greek Fire". But the Crusades help to spread the knowledge of it and we find its use recorded in the *Gesta Dei* at the Siege of Jerusalem. In 1139, an event took place in Western Europe which serves to illustrate the ideas of warfare as developed by Western chivalry. A decision was made by the Second Lateran Council that no machines, "Greek Fire", or similar weapons should be used in Europe against human beings, on account of their being too murderous, a decision, which was in fact, respected for several centuries; a sad commentary on the atavism of modern warfare.

The prevalence of the use of "Greek Fire" crops up frequently in the records of the Crusades (Cinnamus, Jean de Vitry) and its potency, even after the expiration of so many years from its invention, is completely confirmed by crusaders' chronicles such as that of the Joinville, who accompanied Louis IX on his unfortunate sixth Crusade to Damiette (1248). He states that "every man touched by it believed himself lost, every ship attacked devoured by flames."

In 1300 it appears that the secret was known in Europe for a formula for the preparation of "Greek Fire" was included by Marcus Graecus in his treaties on "Weapons for the burning of armies"; the cold barbarism of which title it would be difficult to surpass. He also mentions "liquid fires" consisting of ignited petroleum or resin destillates squirted on to the enemy. The Mongols were also not slow in making use of this refinement of warfare, and Hulagu Khan, a predecessor of the notorius Kublai Khan, arranged in 1253 for a special corps of 1000 men to be brought from Asia Minor to Turkestan and China. This army corps consisted of men specially trained in the use of catapults, grenades and burning naphtha.

The credit for the invention of the best defensive methods against this form of attack appears to have been earned by the Chinese, who learned to protect the roofs of their dwellings with roof-mats of a mixture of rice-straw and grass, coated with clay (1273).

An Arabian book on the art of war (1300) contains the description of a catapult for the projection of jars of burning naphtha, and the advantages of the creation of a special army division of "flingers of naphtha and melting pots" were expounded to such good effect that, according to Ibn Khaldun, the Caliphs adopted the idea and formed a special corps "naffatyn" or "naphtha fire-workers", who were equipped for the more safe execution of their duties with clothes of some fire-proof material (perhaps asbestos), which enabled them to enter the burning city of the enemy. It was not until as late as 1400 that "Greek Fire" began to be rapidly superseded by gunpowder, but already a hundred years later it is used in magicians' tricks only, and the last one mentioning it is the French alchemist Blaise de Vignières (1550) after whose time this weapon which had served so many different peoples for such a long period has fallen into desuetude.

Finally, bitumen was put to curious use in warfare. Bitumen or mastic had long been used for the making of primitive bludgeons or dagger-hafts, as it still is by the Arabs in Northern Syria. During the excavations of Dura Europas, which was besieged and laid in ruins in 265 A.D. by the Persian king Sapor, corridors were found under the old walls (390). After the passages had been dug, the stays were burned with burning faggots, straw, bitumen and sulphur so that the collapse of the passages should make a breach in the wall. Hero of Byzantium stated that wood soaked in bitumen or crude oil was used to set fire to the stays, and the discovery of jars of bitumen bears him out.

Notes

1. EYRINIS D'EYRINIS, *Dissertation sur l'asphalte ou ciment naturel découvert depuis quelques années au Val de Travers dans le comité de Neûchatel, par le Sieur E. d'Eyrinis, professeur grec et docteur en médécine, avec la manière de l'employer tant sur la pierre que sur le bois* (Paris, 1721)
   FORBES, R. J., *Studies in Early Petroleum History* (Leiden, 1958) chap. II
2. SCHOFIELD, M., *The rise and decline of charcoal burning (Science Progress*, vol. XXVI, 1932, 654—661)
   FORBES, R. J., *Metallurgy in Antiquity* (Leiden, 1950, 106)
3. PLINY, *Nat. History* XXXV, 179
4. LIBAVIUS, ANDREAS, *Singularium Andreae Libavii, cont. VIII libros bituminem et affinum historice, physice, chemice; de Petroleis, Ambra, Halosanto, Succino, Gagate, Asphalto, Pissasphalto, Mumia, Lithranthrace* (Frankfort, 1601)
5. —, *Singularium Andreae Libavii. Pars Prima: Liber de Ortu et Natura diatribe* (Frankfort, 1599)
6. ABRAHAM, H., *Asphalt and Allied Substances* (New York, 1929)
7. MARCUSSON, B., *Die natürlichen und künstlichen Asphalte* (2. Auflage, Berlin, 1931)
8. FORBES, R. J., *Short History of the Art of Distillation* (Leiden, 1948)
9. EG, IV, 160. 1
10. Pyr. 51
11. LUCAS, A., *Ancient Egyptian Materials and Industries* (2nd edit. London, 1934, 235—239)
12. SCHÄFER, H., *Aegyptische Inschriften aus dem Kgl. Museum zu Berlin*, vol. I. 22 (Berlin 15004)
13. SETHE, K., *Urkunden* (Leipzig, 1924, vol. I, 145—147)
14. BRUGSCH, H., *Thesaurus* (1181—1182, 11.9—13; 11.2.9)
15. BREADSTED, H., *Ancient Records of Egypt* (3rd impr. Chicago, 1927, vol. IV, 376)
16. SPIEGELBERG, W., *Ein Bruchstück des Bestattungsrituals Apisstiere (Z.f. Aeg. Spr.*, vol. 56, 1920, 1—33)
17. SETHE, K., *Zur Geschichte der Einbalsamierung bei den Aegyptern (Sitz. Preuss. Akad. Wiss.*, 1934)
18. GARDINER, A. H., *Admonitions of an Egyptian Sage* (Leipzig, 1909, iii 7—8)
19. EBBELL, B., *The Papyrus Ebers* (Copenhagen, 1937, p. 67)
20. MENGHIN, O., *Die Ausgrabungen bei Maadi (Mitt. Instit. aeg. Alt.kunde*, vol. 3, p. 151)
21. EG II.684.4
22. BRUGSCH, H., *Hierogl. Demotisches Wörterbuch* (Leipzig, 1880—1882, pag. 679)
23. MASPÉRO, G., *Le Rituel de l'Embaumement* (In: *Mémoires sur quelques Papyrus du Louvre)* (Paris, 1875, page 54)

24. Loret, V., *L'asphalte ou bitume de Judée* (In: *Etudes de droguerie Egyptienne, Rec. des Trav. edit. Maspéro* vol. XVI, 1894, pags. 157—161)
25. Pyr. 754, 879, 2071—2073
26. Chassinat, E., *Le mot mrḥt dans les textes médicaux (Recueil Champollion, Bibl. Ecole Hautes Etudes*, vol. 234, 1922, pags. 447—467)
27. Brugsch, H., *Zwei bilingue Papyri*, vi, 4
28. —, *Hierogl. Demot. Wörterbuch* (Leipzig, 1880—1882, 506)
29. EG II.659.9
30. Maspéro, G., *Le Rituel de l'Embaumement* (In: *Mémoires sur quelques Papyrus du Louvre*, Paris 1875)
31. Loret, V., *L'asphalte ou bitume de Judée (Rec. des Trav. Maspéro*, vol. XVI, 1894, 157)
32. Dioscorides, I. 99
33. Ebbell, B., *The Papyrus Ebers* (Copenhagen, 1937, pag. 67)
34. Fillion, H., *Contribution à la chimie des charbons, pétroles et asphaltes du Liban et de la Syrie (Ann. Fac. Franc. de Méd. de Beyrouth*, 1936, nos. 1 & 2)
    Kelso, J. L. and Powell, A. R., *Glance pitch from Tell Beit Mirsim* (Bull. Amer. School Orient. Res. no. 95, 1944, 14—18)
35. Darmesteter, E., *Etudes Iranniennes*, vol. I, 12—13
36. Herzfeld, E., *Le Mythe Arien du naphte* (II Congrès Mondial du Petrole, Paris, 1937, no. S. 5)
    Brandenstein, W., *Naphtha und Neptunus (Orient. Lit. Z.* vol. 43, 1940, 345—347)
37. CT, iii, 2, 5
38. Hunger, W., *Becherwahrsagung (Leipz. Semit. Studien*, Leipzig, 1903, vol. I, page 80)
39. *Vorderas. Schriftdenkm.*, vi, 228, 3
40. CT., xxxix, 21, 156
41. CT., xxxix, 10, 26
42. CT., xxxix, 10, a. 5
43. Séguin, A., *Etude sur le pétrole dans l'Asie occidentale ancienne. L'Elam et la Mésopotamie* (II Congrès du Petrole, Paris, 1937, nos. R 44, S 5)
44. CT., xxxix, 19, 121
45. AM, 18, 3, 3; BBR. 41, 17
46. or ge_9 (ŠL, 172, 18). See also:
    Boissier, A., *Choix des textes relatifs à la divination assyro-babylonienne* (Paris, 1905—1906, vols i & ii)
    Meissner, B., *Seltene Assyrische Ideogramme* (Leipzig, 1910, No. 11542)
47. Deimel, A., *Sumerische Lexikon* (Rome, 1925, ff)
48. —, *Die Inschriften von Fara. Liste der archaischen Keilschriftzeichen* (Leipzig, 1922, No. 768)
49. Brunnow, J., *A Classified List of all Simple and Compound Ideograms* (Leiden, 1889, nos. 11673 & 11674)
    Delitsch, F., *Sumerisches Glossar* (Leipzig, 1914)
    —, *Assyrisches Handwörterbuch* (Leipzig, 1896)
50. Thompson, R. Campbell, *Dictionary of Assyrian chemistry and geology* (London, 1936)

51. HAUPT, *Beiträge zur Assyriologie* X, ii, 141
ALBRIGHT, W. F., *Classical Weekly*, vol. 31, 1937, No. 2
52. BBR, xxvii, 11
53. LANDSBERGER, B., *Z.f. Assyriologie*, vol. 35, 1924, 233
54. POEBEL, A., *Z.f. Assyriologie*, vol. 39, 1928, 145
55. MEISSNER, B., *Beiträge zum Assyrischen Wörterbuch (Assyrian Studies* No. 6, Chicago, 1932, pags. 4—6)
—, *Seltene Assyrische Ideogramme* (Leipzig, 1910)
56. FORRER, E., *Orient. Lit. Zeitung*, vol. 11, 1937, 673—675
57. CT., vii, 21 a
58. CT, VII, 18390; ITT V, 6978, 6957, 6784, 6878, 8234
REISNER, G., *Tempelurkunden aus Telloh* 13, 121, 1.12
59. STRABO xv, 1; CT, vii, 17775; ITT V, 8234
60. BARTON, G. A., *On Binding-reeds, Bitumen and other Commodities in Ancient Babylonia (J. Amer. Orient. Soc.*, vol. 46, 1926, 297—302)
61. DELITSCH, F., *Assyrisches Handwörterbuch* (Leipzig, 1896, pag. 348, b)
62. MEISSNER, B., *Seltene Assyrische Ideogramme* (Leipzig, 1910, No. 8970)
ITT V, 6957, 6978, 6982, 6784, 6882, 8213; SL 487, 5
63. AM, 38, 6, 7; AM, 13, 4, 4; AM, 93, 1, 11; CT xxiii, 22, 44)
64. HUSSEY, MARY I., *Sumerian Tablet in the Harvard Museum*, vol. ii, 150
ITT V, 755
65. SL, 579
66. SL, 9, 13
67. THUREAU DANGIN, E., *Inscriptions de Sumer et d'Akkad* (Paris 1905, page 157)
ITT, II, 755
68. THUREAU DANGIN, F., see 67
69. FISH, T., *About Building in Ur III (Bull. J. Rylands Libr.*, vol. 18, 1934, No. 1, pags. 134—135)
ITT, V, pag. 1; CT v, 17752; SL 449, 224
70. REISNER, G., *Tempelurkunden aus Telloh*, pags. 121—122
71. THUREAU DANGIN, F., see 67
72. FORRER, E., *Orient. Lit. Zeitung*, vol. 11, 1937, 673—675
73. RTC, 307, r.11.15; ITT V, 6784, 6982, 6892; CT v, Br.M. 17752
74. FISH, T., *About Building in Ur III (Bull. J. Rylands Libr.* vol. 18, 1934, No. 1, pags. 134—135)
75. TC, v, 7, col. i, 1, 3
76. CHIERA, E., *Selected Temple Accounts* (Princeton, 1923, No. 11, col. iii, line 21)
77. SL, 56, 3
78. BARTON, G. A., *J. Amer. Orient. Soc.*, vol. 48, 1926, 297—302
79. ITT, I, 1451, 1, 2; CT v, 38, 213; REISNER, *Tempelurkunden*, 121
80. DRIVER-MILES, *Assyrian Laws*, page 408, line 76 and note on page 481
81. SL, 430, 2 and 461
82. KÖHLER-UNGNAD, *Hammurabi's Gesetz* (Berlin, 1904, vol. III, 128)
83. MEISSNER, B., *Seltene Assyrische Ideogramme* (Leipzig, 1910, No. 1681)
84. BURROWS, M., *Archaic Texts from Ur* (London, 1937, No. 48, ii, 1)
85. MEISSNER, B., *Altbabylonisches Privatrecht*, No. 26, 116

86. —, *Seltene Assyrische Ideogramme* (Leipzig, 1910, No. 8970)
87. SL 488; Delitsch, *Wörterbuch*, pag. 375, ii
88. Barenton, H. de, *De l'origine des langues* (Paris, 1923, vol. i, pag. 293)
89. Strassmaier, J. N., *Die Inschriften des Cambyses* (Leipzig 1887)
90. Thucydides, IV, 100
91. Theophrastus, *de Lapid.* 16
92. Dioscorides, V. 181; Strabo, VII, cap. 316
93. Dioscorides, V. 146; Pliny, *Nat. Hist.* XXXVI, 141
94. Cunnington, *Antiquity*, vol. VII, 1933, page 89
Davies, O., *Roman Mines in Europe* (Oxford, 1935, 153)
95. Thompson, R. Campbell, *Dictionary of Assyrian Chemistry and Geology* (London, 1936)
96. Strassmaier, J. N., *Die Inschriften des Cambyses* (Leipzig, 1887, 105, 1)
97. Guthe, H., *Bibelwörterbuch* (Leipzig, 1903)
98. Theophrastus, *Peri Lithoon*, cap. 16 & 35 (German transl. by K. Mieleitner: *Geschichte der Mineralogie im Altertum und Mittelalter, Fortschritte der Mineralogie* 1922, page 427)
Forbes, R. J., *Studies in Ancient Technology* Vol. VII (Leiden, 1963)
99. Imbert, P., *Ueber Salzbrunnen in China (Dinglers Polyt. J.*, vol. xxxiv, 1828, pag. 72) (Original in *Ann. Assoc. Propag. de la Foi*, vol. III, 1828, page 369)
Heim, A., *Das älteste Bohrfeld der Welt in Szechuan (China) (Petroleum*, vol. xxvi, 1930, pags. 171—173)
Forbes, R. J., *Studies in Early Petroleum History* (Brill, Leiden, 1958, pag. 173—181)
100. Séguin, A., *Etude sur le pétrole dans l'antiquité égyptienne* (II Congrès Mondial du Pétrole, Paris, 1937, R.43, S.5)
101. Schoo, J., *Romeinsche aardoliewinning aan de golf van Suez (Tijdschr. Ned. Aardr. Gen.*, vol. 51, 1934, 883—888)
Deines-Grapow, *Wörterbuch der ägyptischen Drogennamen* (Berlin, 1959 250)
102. Dioscorides, V. 158
103. Joachim, H., *Der Papyrus Ebers* (Berlin, 1905), but see Harris, J. H., *Lexicographical Studies in Ancient Egyptian Minerals* (Berlin, 1961, page 25)
104. Strabo, XVI, cap. 764
105. Diodor, XIX, 98
106. Vitruv, VIII.3.8
107. Aristotle, *Mirab.* 113
108. Garstang, J., *The fate of Jericho as revealed by the spade (Ill. London News*, 16-12-1933)
109. Albright, W. F., *Classical Weekly*, vol. 31, 1937, No. 2
110. Vitruv, VIII.3.9
111. Pliny, *Nat. Hist.* XXXV.178
112. Diodor, XIX, 98.2
113. Diodor, II.48.6
114. Josephus, *Bell. Iud.* IV, 476

115. STRABO, XVI cap. 763
116. SUIDAS, *Bibliographica* (edit. ADLER, page 396)
117. ISIDORE OF SEVILLE, *Originum sive etymologiarum libri XX*, (edit. Lindsay 1911, Lib. XVI, cap. 2 & 4)
118. Gen. XIV, 10
119. LUCAS, A., *Preservative Materials used by the Ancient Egyptians in Embalming (J. Egypt. Arch.*, vol. I, 1914, 241)
—, *Ancient Egyptian Materials and Industries*, London, 1962)
120. DIODOR, XIX, 98
121. TARN, W. W., *Cambridge Ancient History*, vol. X, 1934, page 67
122. ISIDORE, *Etym.* XVI.2
123. VITRUV, VII.3.8
124. DIOSCORIDES, I.83
125. PLINY, *Nat. Hist.* XXXV.178
126. DUSSAUD, R., *Byblos et la mention des Giblites dans l'ancien Testament (Syria*, vol. 4, 1926, page 310)
127. CLARIS, P., *Analyse des résidues trouvés dans le grand sarcophage de Byblos (Syria*, vol. 4, 1926, 79—80)
128. SÉGUIN, A., *Etude sur le pétrole dans quelques pays de l'Orient ancien.* - I. *Canaan et la Phénicie* (II Congrès du Petrole, Paris, 1937, R.45, S.5)
129. MORET, A., *Histoire de l'Orient* (Paris, 1926, page 299)
130. VITRUVIUS, VIII.3.8
131. THEOPHRASTUS, *de Lapid.* 85
132. DIOSCORIDES, V.181
133. ORIBASIUS, Book 13
134. AETIUS, II.9
135. GALEN, *De Fac. Simpl. Med.* (edit. KUHN) IX.4
136. STRABO, VII, cap. 316
137. DIOSCORIDES, V. 146
138. PLINY, *Nat. Hist.* XXXVI, 141
139. SOLINUS, XXII, 11
140. ISIDORE, XVI, 4.3
141. BAILEY, K. C., *The Elder Pliny's Chapters on Chemical Subjects* (London, 2 vols., 1929 & 1932)
142. EPIPHANIOS, *Panarion* (edit. HOLL, Leipzig, 1915)
143. VORAGINE, J. DE, *The Golden Legend* (edit. DENT, vol. II, page 117)
144. STAHL, A. F. VON, *Das Erdölvorkommen in Türkisch-Armenien (Petroleum Z*, vol. XXIII, page 115)
145. PLINY, *Nat. Hist.* II.235
146. Sabbat 26a
147. MEHREN, A. F., *Manuel de la Cosmographie du Moyen Age* (transl. of DIMASHQÎ) (Notes on asphalt and mumiya pages 69—97, asphalt seepages near Hit see page 153)
SCHWEER, WALTER, *Die einheimschen Erdölindustrie in Mesopotamien und Persien (Petroleum*, vol. XV, 1920, No. 10, pags. 397—399)
—, *Die türkisch-persischen Erdölvorkommen (Abh. Hamburger Kolonialinstitutes*, Band XL, 1919)
148. ANDRAE, W., *Mitt. D. Or. Ges.*, vol. XXII, page 64)

149. Ammian, *Res Gestae* XXIII.6.15
150. Smith, G., *Assyrian Discoveries* (London, 1876, page 96)
151. Ibn Jubayr, *The Travels of Ibn Jubayr*, edited by William Wright, second edition revised by M. J. de Goeje, *Gibb Memorial Series*, vol. V, London, 1907, pags. 233—234
   Forbes, R. J., *Studies in Early Petroleum History* (Leiden, 1958, pags. 149—162)
152. Gadd, C. J., *Cuneiform Texts from Babylonian Tablets* XXXIX, 22.10
153. *Ill. London News*, Jan. 19, 1935, page 82
154. Thureau Dangin, F., *Die Sumerischen und Akkadischen Königsinschriften* (Leipzig, 1907, page 106, xvi. 9)
155. Ebeling, Meissner und Weidner, *Die Inschriften der alt-assyrischen Könige*, pags. 74 & 77
156. Luckenbill, D., *The Annals of Sennacherib* (Chicago, 1924, page 105, lines 88 ff)
157. *Beiträge zur Assyriologie*, vol. iii, pag. 323, col. 5, line 22
158. Séguin, A., *Etude sur le pétrole dans quelques pays de l'Orient ancien* - I. *Canaan et la Phénicie* (II Congrès Mondial du Pétrole Paris, 1937, R. 45, S.5)
159. Vitruv, VIII.3.8
160. Dioscorides, I.83
161. Diodor, II.12
162. Strabo, XVI, cap. 743
163. Hackford, Lawson & Spielmann, *On an Asphalt Ring from Ur of the Chaldees (J. Inst. Petr. Techn.*, December 1931, page 738)
164. Forbes, R. J., *Note on a lump of asphalt from Ur (J. Inst. Petr. Techn.*, vol. XXII, 1936, page 180)
165. Meissner, B., *Babylonien und Assyrien* (Heidelberg, 2 vols, 1920—1925)
166. Landsberger, B., *Ueber die Völker Vorderasiens im dritten Jahrtausend v. Chr. (Z.f. Assyriologie*, vol. 35, 1924, pag. 39—42)
167. Rawlinson, H. C., *Inscriptions of Western Asia* IV², 1892, 59, 7b
   —, *History of Herodotus* (London, 1858, vol. I, page 316 note)
168. Sethe, K., *Urkunden*, vol. IV, page 708
   Forbes, R. J., *Studies in Ancient Technology vol. IX* (Leiden, 1964, p. 53)
169. Herzfeld, E., *Masjid-i-Suleiman (Naft Magazine*, vol. V, 1929, pag. 5—9)
170. —, *Archaeological History of Iran* (London, 1936, page 93)
171. Cook, A. B. & Sheppard, C., *Historical Records relating to Oil (J. Inst. Petr. Techn.*, 1927-28, pag. 124—134)
172. Schröder, O., *Keilschrifttexte aus Assur verschiedenen Inhalts* xiv, Leipzig, 1920
173. Scheil, V. & Gauthier, A., *Annales de Tukulti Ninip II* (Paris, 1909, No. 178)
174. Herodotus, I.179
175. Langdon, St., *Die neubabylonischen Königinschriften*
   Strassmaier, J. N., *Die Inschriften des Nabuchodonosor* (Leipzig, 1887)
   —, *Die Inschriften des Nabonidos* (Leipzig, 1887)
176. —, *Die Inschriften des Cambyses* (Leipzig, 1887)

177. CLAY, J., *Neo-Babylonian Letters from Erech* (New Haven, 1919, Nos. 98 & 161)
178. *Amer. J. Sem. Lang.*, vol. 32, No. 84
179. THOMPSON, R. CAMPBELL, *Cuneiform Texts* (London) xxii, No. 1, 84
180. See ITT, vol. V
181. PHILOSTRATOS, *Vit. Apoll.* I, 24
182. PLINY, *Nat. Hist.* I. 99
183. PLINY, *Nat. Hist.* VI. 99
184. ARISTOTLE, *De Mir. Ausc.*, cap. 35, 41, 113, 116 & 127
185. PLINY, *Nat. Hist.* XIV, 128
186. —, *Nat. Hist.* XXXV, 180
187. —, *Nat. Hist.* II, 235
188. SEYBOLT, R. F., *A troublesome Greek word mydiacon (Speculum*, vol. 21, 1946, 38—41)
189. SCHWARZ, P., *Iran im Mittelalter nach den arabischen Geographen* (Naft see vii 419 & 869, mumiya II, 95)
190. MARCO POLO, *Book of Travels* (edit. DENT, Book I, cap. IV & XI)
FORBES, R. J., *Studies in Early Petroleum History* (Leiden, 1958, pags. 154—162)
191. PLINY, *Nat. Hist.* II, 235
192. PLUTARCH, *Alexander*, cap. 35—37
193. STRABO, XI, cap. 518
194. RUSKA, J., *Das Steinbuch aus der Kosmographie des Zakarija ibn Muhammed ibn Mahmud al-Kazwînî* (Beilage zum *Jahresb. Oberrealschule Heidelberg;* pag. 43—44 on alkîr, naphtha and mumiya)
195. VITRUV, VIII.3.8
196. DIOSCORIDES, I.83
197. HERODOTUS, IV. 195
198. DIOSCORIDES, I.83
199. VITRUV, VIII.3.8
200. AELIAN, *Variae historiae Lib. XIII*, cap. 16 (edit. LÜNEMANN, 1811)
201. DIOSCORIDES, I.84
202. STRABO, VII. cap. 316
203. POQUEVILLE, H., *Voyage dans la Grèce*, vol. I, page 271
204. THEOPHRASTUS, *De Lapid.* cap. 35
205. DIOSCORIDES, I.83
206. PLINY, *Nat. Hist.* XXXV, 179
207. THEOPHRASTUS, *De Lapid.* 16
208. KELLER, R., *Pfahlbauten* (8. Bericht, 1879)
VOUGA, P., *The oldest Swiss Lake Dwellings (Antiquity*, vol. II, 1928, page 387)
FORBES, R. J., *Studies in Early Petroleum History* (Leiden, 1958, Chap. II)
209. TACITUS, *Hist.* V. 6
210. STRABO, XVI, cap. 764
211. HERODOTUS, VI. 119
212. HERODOTUS, IV. 195
213. PLINY, *Nat. Hist.* XXXV. 179

214. Agricola, *De Re Metallica* 7, 4
215. Pliny, *Nat. Hist.* XXXV, 178
216. Pliny, *Nat. Hist.* XV, 8; XXIV, 24
217. Dioscorides, I.84893
218. Ruska, J., *Das Steinbuch des al-Kazwînî* (see 194)
219. Forbes, R. J., *Short History of the Art of Distillation* (Leiden, 1948)
220. Agricola, G., *Zwölf Bücher vom Berg- und Hüttenwesen* (Berlin, 1929, Book XII)
221. —, *De natura fossilium* (1546)
—, *Opera* 1657, pag. 595
222. Pliny, *Nat. Hist.* XXXV, 194
223. Strabo, VII, cap. 316
224. Pliny, *Nat. Hist.* XXIV, 41
225. Glotz, G., *L'histoire de Délos d'après les prix d'une denrée.* (*Revue des Stud. Grecques*, vol. 29, 1916, pag. 281)
226. Theophrastius, *Bot.* IX. 3. 1
227. Pliny, *Nat. Hist.* XVI, 52
228. Weiker, W., *Ein alter Holzteerofen Thüringens* (*Technikgeschichte*, Band 23, 1934, pag. 134)
229. Pliny, *Nat. Hist.* XIV, 122 & 127; XVI, 38—52
230. Strabo, III. cap. 144
231. —, IV, cap. 207
232. —, V, cap. 218
233. —, XI, cap. 498
234. Schofield, M., *The rise and decline of charcoal burning* (*Science Progress*, vol. XXVI, 1932, pag. 654—661)
235. Pliny, *Nat. Hist.* XIV, 134; XV, 62
236. —, *Nat. Hist.* XVI, 56 & 158
237. —, *Nat. Hist.* XXXVI, 166
238. —, *Nat. Hist.* XXXV, 41
239. Homer, *Iliad* XI. 533
240. Lucian, *Iup. trag.* 33
241. Watelin, A., *Rapport des fouilles de Kish* (*J. Asiatique*, 1929, pag. 107—118)
242. King, L. W., *History of Sumer and Akkad* (London, 1916, pag. 166)
243. DP. xii, 20, 62 & 162
Contenau, G., *Manuel de l'Archéologie Orientale* (Paris, 1927, vol. II, page 807)
244. Handcock, S. P., *Mesopotamian Archeology* (London, 1912)
245. Mercier, M., *Quelques points de l'histoire du pétrole.* (*Bull. No. 30 Assoc. Franc. Techn. Pétrole*, 1937)
246. Thureau Dangin, F., *Textes mathématiques babyloniens* (*Revue d'Assyr.*, vol. XXXIII, 1936, pag. 79)
247. Philostratus, *Vita Apoll.* III, 1
248. Strabo, XVI, cap. 764
249. —, VII, cap. 316
250. Pliny, *Nat. Hist.* XXXV, 179
251. Isidore, *Etym.* XVI. 2

252. Pliny, *Nat. Hist.* XXXV, 180
253. Dioscorides, I. 99
254. Pliny, *Nat. Hist.* VII, 65
255. *Eccl.* XIII, 1
256. Tacitus, *Hist.* V, 6
257. Pliny, *Nat. Hist.* XVI, 22
258. Strabo, XVI, cap. 768.
259. Budge, E. A. Wallis, *The Mummy* (London, 1925)
Forbes, R. J., *Studies in Early Petroleum History* (Leiden, 1958, pags. 162—167)
260. Marshall, Sir J., *Mohenjo Daro and the Indus Civilisation* (London, 1931)
—, *Indus Civilisation* (In: *Ann. Rep. Archaeol. Survey India* 1928-24 and 1925-26)
Mackay, E., *Early Indus Civilisations* (London, 1948)
261. Forbes, R. J., *Untersuchungen betreffend die ältesten Anwendungen von Bitumen in Mesopotamien (Bitumen*, 1935, No. 1, page 9)
262. Willocks, Sir W., *The irrigation of Mesopotamia* (London 1911, page 89)
263. Scheidig, A., *Der Loess und seine geotechnischen Eigenschaften* (Leipzig, 1934)
264. Meissner, B., *Babylonien und Assyrien* (Heidelberg, 1920), vol. I
265. Banks, E., *Bismaya* (London, 1912, pags. 191, 271 & 314)
266. Thompson, R. Campbell, *Dictionary of Assyrian Chemistry and Geology* (London, 1936)
267. Meissner, B., *Babylonien und Assyrien* (Heidelberg 1920, vol. I)
268. *Rev. d'Assyr.*, vol. XXXIII, page 87
269. Parkhurst, R., *Assyrian Engineering (Civil Engineering*, 1932, No. 6, page 345)
270. Helfritz, H., *Das Chicago der Wüste* (Berlin, 1932)
271. Goldschmidt, V. M., *Undersökelser over Lersedimenter (Beretning om Nordiske Jordbrugforsker*, *Nordiske Jordbrugforkning*, 1926, page 434)
272. Strassmaier, J. N., *Inschriften des Nabonidus* (Leipzig, 1889, pag. 231)
—, *Inschriften des Kyrus* (Leipzig, 1890, page 277)
273. Thompson, R. Campbell, *Dictionary of Assyrian Botany* (London, 1949)
274. Forbes, R. J., *Untersuchungen betreffend die ältesten Anwendungen von Bitumen in Mesopotamien (Bitumen*, 1935, No. 1, page 9)
275. Fish, T., *About Building in Ur III (Bull. J. Rylands Libr.* vol. XVIII, 1934, No. 1, 134—135)
276. Delougaz, P., *Plano-convex bricks and their method of employment (Studies in Ancient Oriental Civilisation*, No. 7, Chicago, 1933)
277. Rathgen, B., *Chemiker Ztg.* 1913, page 441
278. Xenophon, *Anabasis* II. 4. 12
279. Vitruv, I. 5. 8
280. Strabo, XVI, cap. 738
281. Cassius Dio, *Roman History LXVIII*, 27 (edit. *Loeb. Class. Library*)
282. Meissner, B., *Babylonien und Assyrien* (Heidelberg 1920, vol. II)

283. Fish, T., *About building in Ur III (Bull. J. Rylands Libr.* vol. XVIII, 1934, pag. 134—135)
284. Reuther, O., *Die Innenstadt von Babylon* (Leipzig, 1926)
285. Herodotus, I. 179
286. Jordan, J., *Ausgrabungen in Uruk 1930—31 (Abh. Preuss. Akad. Wiss. Phil. Hist. Klasse*, 1932)
287. Forbes, R. J., *Untersuchungen betreffend die ältesten Anwendungen von Bitumen in Mesopotamien (Bitumen*, 1935, No. 1, page 9)
Nellensteyn, F. J., *Onderzoekingen over asphalt gevonden bij de opgravingen te Mohenjo Daro (Chem. Weekblad*, 1934, page 543)
(German transl. *Bitumen* 1935, page 15)
288. Bonomi, J., *Niniveh and its Palaces* (London 1852)
289. Banks, E., *Bismaya* (London, 1912, pag. 191, 271 & 314)
290. Andrae, W., *Die Festungswerke Assur (230 Wiss. Veröff. Dtsche Orient Gesch.*, 1913)
291. —, *Mitt. D. Orient. Ges.*, vol. XXII, page 33
292. Meissner, B., *Die Bauinschriften des Sanheribs* (Berlin, page 6)
293. Koldewey, R., *Mitt. D. Orient. Ges.*, vol. XXXIII, page 9
294. Herodotus, I. 186
295. Koldewey, R., *Mitt. D. Orient. Ges.*, vol. XXXIII, page 8
296. —, *Excavations at Babylon* (London, 1914, pag. 31, 54, 113 & 168)
297. Reuther, O., *Die Innenstadt von Babylon* (Leipzig, 1926)
298. Andrae, W., *Mitt. D. Orient. Ges.*, vol. XXV, page 4
299. Koldewey, R., *Excavations at Babylon* (London, 1914)
300. Andrae, W., *Alte Feststrassen im Nahen Osten* (Leipzig, 1941)
301. Forbes, R. J., *Ancient Roads and their construction* (Amsterdam, 1934) (Reprint, Amsterdam, 1964)
302. Colberg, O., *Die Isolierung von Maschinengründungen gegen Ausbreitung der Erschütterungen und Schwingungen (Bitumen*, 1934, No. 10, pag. 251)
303. *Mitt. Vorderas. Ges.*, vol. X, page 312
Forbes, R. J., *Studies in Ancient Technology* vol. VI (Leiden, 1958)
304. Condinus, Georgius, *Description of Byzantium* (Transl. in *Quellen zur Byz. Kunstgeschichte* 1878, page 275)
305. Strabo, XVI, cap. 739
306. Reuther, O., *Die Innenstadt von Babylon* (Leipzig, 1926)
307. Koldewey, R., *Excavations at Babylon* (London, 1914)
308. Lucas, A., *Ancient Egyptian Materials and Industries* (London, 1962)
309. Pliny, *Nat. Hist.* XXXIV, 15
310. —, *Nat. Hist.* XXXV, 182
311. Lippmann, E. O. von, *Entstehung und Ausbreitung der Alchemie* (Berlin 1919, vol. I, page 12)
312. Forbes, R. J., *Short History of the Art of Distillation* (Leiden, 1948)
313. Haas, P., *Chemical News*, 17-12-1909 (See also 378)
314. Newton Friend, J., *An Introduction to the Chemistry of Paints* (London, 1910, page 95)
315. Strabo, XVI, cap. 740
316. Reuhter, O., *Die Innenstadt von Babylon* (Leipzig, 1926)

317. MACKENZIE, D. A., *Myths of Babylonia and Assyria* (London, 1916, page 191)
ROGERS, B. W., *Cuneiform Parallels* (London, 1912, pag. 91 & 135)
318. EXODUS II. 2
319. HERODOTUS, I. 194; FORBES, R. J., *Asfalt voor het biezen kistje* (*Olie*, Dec. 1961, pag. 369—372)
320. HEIDEL, A., *The Gilgamesh Epic and Old Testament Parallels* (Chicago, 1946)
321. *Genesis* VI. 4
321a. SALONEN, A., *Die Wasserfahrzeuge in Babylonien* (Helsingfors, 1939; SALONEN, A., *Nautica Babylonica* (Helsinki, 1942); DRIVER & MILES, *Bab. Laws II*, pp. 257
322. TROELTSCH, H. VON, *Die Pfahlbauten des Bodensees* (1902, p. 54 & 66)
VOUGA, P., *The oldest Swiss Lakedwellings* (*Antiquity*, vol. II, 1928, page 387)
323. WOOLLEY, SIR L., *Excavations at Ur* (*Ant. J.*, vol. 8, 1928, page 1)
HALL & WOOLLEY, *Ur Excavations* (London, 1927, vol. I)
LANGDON, ST., *Excavations at Kish* (Paris, 1924, 4 vols.)
—, *Ausgrabungen in Mesopotamien seit 1918* (*Der Alte Orient*, vol. 26, 1928, No. 1)
324. CONTENAU, G., *Manuel d'Archéologie Orientale* (Paris, 1927, vol. I)
325. MEQUENEM, R. DE, *Contribution à l'étude des outils en pierre trouvés dans les ruines de Suse* (*Anthropologie*, 1923, page 33)
326. DP vol. I, page 66
327. BANKS, E., *Bismaya* (London, 1912)
328. WIJNGAARDEN, W. D. VAN, *Het Egyptische Mummieportret* (*Elsevier's Maandschrift*, vol. VII, 1933, page 16)
329. FRANKFORT, H., *The Iraq Excavations of the Oriental Institute 1932—1933* (*Orient. Instit. Comm.* No. 17, Chicago, 1934)
330. LUCIAN, ALEXANDER, (edit. LANGENSCHEIDT, 1867, cap. 22)
331. HALL, H. R. H., *J. Egypt. Archeol.*, vol. 8, 1922, page 247
332. WOOLLEY, SIR L., *Digging up the past* (London, 1930)
333. SARZEC, H. DE & HEUZEY, *Découvertes en Chaldée* (Paris, 1884—1912, 2 vols.)
HANDCOCK, S. P., *Mesopotamian Archaeology* (London 1912)
334. RAWLINSON, G., *The Five Great Monarchies of the Ancient Eastern World* (London, 1875, vol. I, page 367)
335. K. GSELL, *Eisen, Bronze und Kupfer bei den alten Aegyptern* (Karlsruhe, 1910, page 79)
336. HACKFORD, LAWSON & SPIELMANN, *On an Asphalt Ring from Ur of the Chaldees* (*J. Inst. Petr. Techn.*, 1931, page 738)
CALKIN, B. J., *Kimmeridge coal money, the Romano-British shale armlet industry* (Proc. Dorset nat. hist. & arch. soc. 75, 1955, 45—71)
LIVERSIDGE, J., *Furniture in Roman Britain* (London, 1955)
337. AM, 93. 11
338. EPPING-STRASSMAIER, *Z.f. Assyr.*, vol. 6, page 242, line 15
339. MEIER, G., *Die Assyrische Beschwörungsserie Maqlû* (*Archiv f. Orientforschung*, Berlin, 1937, Beiheft 2)

340. CT. xvi, 22, 304
341. KARI, No. 473, obv. 9
342. KARI, No. 69, r
343. CT, xvii, 25, 26
344. CT, xxxix, 19, 121
Boissier, A., *Choix des textes relatifs à la divination assyro-babylonienne* (Paris, 1905-06, 2 vols)
345. CT, xxxix, 22, 10
346. CT, iii, 2, 5
347. Meissner, B., *Babylonien und Assyrien* (Heidelberg 1925, vol. II)
348. Langdon, St., *A Babylonian Ritual of Sympathetic Magic by Burning Images (Revue d'Assyr*, vol. 26, 1929, pag. 39—42)
349. Strabo, VII, 5, 8
Cook, A. B. & Sheppard, C., *Historical Records relating to Oil (J. Inst. Petr. Techn.*, 1927-28, p. 124—134)
350. KARI No. 1, V. 76
351. AM, 18. 3. 3; 73. 11. 8
352. BBR, 41. 17
353. AM. 78. 10. 4
354. AM 15. 3. r. 7; 70. 3. 1
355. AM 38. 6. 7
356. AM 15. 3; 73. 1. 7
357. KARI, No. 473, 10
358. AM 90. 1
359. Gruppe, O., *Griechische Mythologie und Religionsgeschichte (Handb. klass. Altert. Wiss.*, München, 1906, vol. V, ii, page 889, note 4)
360. Pliny, *Nat. Hist.* XXXV, 180 & 182
361. Dioscorides, I. 85
362. Josephus, *Bell. Iud.* IV. 481
363. Varro, Lib. I. cap. 12
364. Virgil, *Georg.* III. 451
365. *Geoponica* XVII. 16. 1
366. Philostratus, *Vit. Apoll.* I. 24
367. Pliny, *Nat. Hist.* XIV. 20
368. Aetius, *Tetrabiblos* I. 2. 49
369. *Geoponica* XIV. 11. 4
370. —, XIII. 10. 7
371. —, XVII. 16
372. —, XIV. 11
373. Reutter de Rosemont, L., *De l'embaumement avant et après Jésus Christ* (Paris, 1912)
374. Tschirsch, A., *Ueber die im ersten Jahrtausend v. Chr. bei der Einbalsamierung in Aegypten und Karthago benutzten Harze (Archiv f. Pharmazie*, 1912, page 170)
375. Budge, E. A. Wallis, *The mummy* (London, 1925)
376. Lucas, A., *Preservative Materials used by the Ancient Egyptians in Embalming (J. Egypt. Archaeol.*, vol. I, 1914, page 132)
377. Coremans, P., *Chronique d'Egypte*, No. 24, 1937

378. BUDGE, E. A. WALLIS, *The Mummy* (London, 1925)
FORBES, R. J., *Studies in Early Petroleum History* (Brill, Leiden, 1958) pags. 162-167
PROOST, W., *Mumia (J. de Pharmacie de Belgique* 1954, pags. 47-53)
379. DIODOR, XIX, 6
380. STRABO, XVI, cap. 764
381. LUCAS, A., *Ancient Egyptian Materials and Industries* (London, 1962)
382. SPIELMANN, P. E. *To what extent did the ancient Egyptians employ Bitumen for Embalming (J. Egypt. Archaeol.* vol. 18, 1933, page 177)
383. MÖLLER, G., *Die beiden Totenpapyri Rhind* (Berlin, 1913, I. 3, demot. 8, hierat. 9)
384. THUCYDIDES, II. 77
385. —, IV. 100
386. PLINY, *Nat. Hist.* II. 104
387. PHILOSTRATOS, *Vita Apoll.* III. 1
388. BERTHELOT, M., *Les compositions incendaires dans l'Antiquité et au Moyen Age (Revue des Deux Mondes*, vol. 106, 1891, page 787)
DIELS, H., *Antike Technik* (Leipzig, 1924, 3. Aufl. pag. 108-114)
HULME, E. WYNDHAM, *German wild-fire trials at Oxford (Trans. Newcomen Soc.*, vol. X, 1929-30, pag. 89-99)
LIPPMANN, E. O. VON, *Zur Geschichte des Schiesspulvers (Abh. und Vorträge z. Gesch. d. Naturwiss.*, 1906, page 125)
ZENGHELIS, C., *Le feu grégois et les armes à feu des Byzantins (Byzantion*, vol. VII, 1932, pag. 265-286)
MERCIER, M., *Le Feu grégois* (Paris, 1952)
BROCK, A. ST. H., *A History of Fireworks* (London, 1949)
PARTINGTON, J. R., *A History of Greek Fire and Gunpowder* (Heffer, Cambridge, 1960)
FORBES, R. J., *More Studies in Early Petroleum History* (Brill, Leiden, 1959) chap. IV
389. LIVY, XXXIX. 13
390. MESNIL DU BUISSON, F. DU, *Les fouilles de Doura-Europos et le naphte* (II. Congrès Mondial du Petrole, Paris, 1937, S. 5)

ABBREVIATIONS

AM.: R. CAMPBELL THOMPSON, *Assyrian Medical Texts* (Oxford, 1923)
BBR.: H. ZIMMERN, *Beiträge zur Kenntnis der Babylonischen Religion* (Leipzig, 1896—1900)
CT.: *Cuneiform Texts from Babylonian Tablets in the British Museum* (London)
DP.: *Mémoires de la Délégation en Perse* (Paris)
EG.: A. ERMAN & H. GRAPOW, *Wörterbuch der aegyptischen Sprache* (Leipzig, 1926—1931, 5 vols.)
ITT.: H. DE GENOUILLAC, *Inventaire des Tablettes de Telloh* (Paris, vol. II, 1910, vol. V, 1921)
KARI.: E. EBELING, *Keilschrifttexte aus Assur religiösen Inhalts* (Leipzig, 1919—1920, 2 vols.)

Pyr.: C. H. Sethe, *Die altaegyptischen Pyramidentexte* (Leipzig, 1908—1922, 4 vols.)
ŠL.: A. Deimel, *Sumerisches Lexikon* (Rome, 1925—1937, 3 vols.)

Bibliography

Abraham, H., *Asphalt and allied Substances* (New York, 1945, 5th edit.)

Blümner, H., *Technologie und Terminologie der Gewerbe und Künste bei den Griechen und den Römern* (Leipzig, 1879, vol. II, pag. 347)

Chapot, V., *Le bitume dans l'Antiquité (J. des Savants*, 1939, pags. 127-132)

Forbes, R. J., *Untersuchungen betreffend die ältesten Anwendungen von Bitumen in Mesopotamien (Bitumen*, 1935, Heft 1-3)

—, *The Nomenclature of Bitumen, Petroleum, Tar and Allied Products in Antiquity (Mnemosyne, Bibl. Class. Bat.*, 3e Sér. 4, 1936)

—, *Bitumen and Petroleum in Antiquity*, Leyden, 1936

—, *Bericht über einen bei Ur gefundenen Asphaltklumpen (Asphalt und Teer*, vol. 38, 1938, No. 18, pags. 257-258)

—, *Bitumen and Petroleum in Antiquity, new notes (Ambix*, vol. II, 1938, pags. 68-92)

—, *Petroleum and Bitumen in Antiquity (J. Instit. Petr.*, vol. XXV, 1939, pags. 19-23)

Forbes, R. J., *Studies in Early Petroleum History* (Brill, Leiden, 1958)

Forbes, R. J., *More Studies in Early Petroleum History* (Brill, Leiden, 1959)

Griffiths, J. G. A., *Resin and Pitch from ancient Egyptian Tombs (Analyst*, vol. 62, 1937, pags. 703-709)

Hecht, H., *Erdölvorkommen im Altertum (Oel und Kohle*, vol. 38, 1942, No. 43, pags. 1303-1306)

Lockhart, L., *Iranian Petroleum in Ancient and Medieval Times (J. Inst. Petr.*, vol. XXV, 1939, No. 183, pags. 1-18)

Lucas, A., *Preservative Materials used by the Ancient Egyptians in Embalming* (Cairo, 1911)

Meier, K., *Ueber die echte Mumie (Archiv. f. Gesch. d. Medizin*, vol. 30, 1937, pags. 62-77)

Mercier, Maurice, *Quelques points de l'histoire du pétrole (Bulletin No. 39 de l'Association Française Techniciens du Pétrole*, 1937: II Congrès Mondial du Petrole, S. 5)

—, *Le bitume a-t-il été utilisé dans la haute Antiquité egyptienne? Thalès*, vol. III, 1936, pags. 68-69)

—, et Séguin, A., *L'exode témoigne-t-il de la présence du pétrole? (La Revue Pétrolifère*, No. 779, 1938)

—, et Séguin, A., *L'Epigraphie et les fontaines ardentes du Dauphiné (Bull. Assoc. Franc. Techn. Pétrole*, 1939, Nos. 48 & 49)

Nakhai, M., *Le pétrole en Iran* (Paris, 1938)

Nellesteyn, F. J. en Brand, J., *Onderzoekingen over asfalt, gevonden bij Mesopotamische opgravingen (Chemisch Weekblad*, vol. 33, 1936, pag. 261-263)

Séguin, A., *Recherches sur le pétrole dans l'antiquité (Revue des Questions Historiques*, Janv. 1936, 3)

—, *Etude sur le pétrole dans l'antiquité égyptienne.* (II Congrès Mondial du Pétrole, Paris, 1937, R. 43, S. 5)

—, *Etude sur le pétrole dans quelques pays de l'Orient ancien.* I. *Canaan et la Phénicie* (II Congrès Mondial du Pétrole, Paris, 1937, R. 45. S. 5)

—, *Etude sur le pétrole dans l'Asie Occidentale Ancienne. L'Elam et la Mésopotamie* (II Congrès Mondial du Petrole, Paris, 1937, R. 44, S. 5)

—, *Etude sur le pétrole dans l'Antiquité grecque et latine (Revue des Questions Historiques,* 1938, pags. 36-71)

TOUTAIN, J., *Sur l'ancienneté des gisements de pétrole de la région de Mossoul (Bulletin Société Sciences historiques et naturelles Semur- en Auxois,* 1926, No. 4, 3—4)

UGNAD, A., *Das Gilgamesch Epos* (Göttingen, 1911, 31, 49, 51, 54, 194, 215)

WOOG, P., *Note sur la recherche du bitume sans la matière dont sont constitués deux Oushabti (Thalès,* vol. III, 1936, pags. 70-82)

Chapter II

THE ORIGIN OF ALCHEMY

The furnace proveth the potter's vessels
so the trial of man is in his reasoning.
Eccles. XXVII. 5

Introduction

The evolution of chemistry, though still shrouded in mystery, is a problem that should intrigue every student of the history of science for it presents us with the unique opportunity of studying the birth of a new science. Where the origins of mathematics and astronomy are forever veiled in darkness, because we are unable to study the world of the spirit in prehistory, chemistry arose in historical times.

This involves a review of our attitude towards alchemy. The contempt with which former generations treated alchemy was both unfair and unjustified. As more evidence was recovered concerning the background and the evolution of alchemy we were able to judge the Art more objectively. We now realize that those who strove to attain knowledge of the Absolute by means of the Philosopher's Stone were no less scientists that we who look for the truth behind the phenomena of the physical world in our mathematical logical way.

Alchemy can in truth be called a science, for though its structure is not mathematical it describes, classifies and draws conclusions from analogies. It combines both theoretical speculation and empirical techniques, the latter factor occuring much more frequently in alchemy than in any other ancient field of science. Nor does it differ profoundly from modern chemistry in its aims. It propounds theories about the elements and their combination in the different artificial and natural compounds, it discusses the changes which these elements undergo, the changes which reacting compounds undergo and it collects empirical facts of the technology of such compounds. It certainly differs from chemistry in method and in the fact that it applies philosophical and religious tenets to a field of natural science. We may in fact describe alchemy as the early, qualitative phase of chemistry.

Origin of the word alchemy

Most textbooks will agree that alchemy was born in Egypt and this is in agreement with Greek tradition which ascribed in hoary age and superhuman wisdom to everything Egyptian. Certainly the oldest documents on alchemy show us an Egyptian school of alchemists flourishing in the shadows of the great Academy of Alexandria in Hellenistic times from the first century A.D. onwards. They seem to have derived the impulse for their researches from Bolos of Mendes, the Neo-Pythagorean who flourished in the second century B.C. in Alexandria.

The term alchemy, derived from the Arabic "al-kîmîya", goes back to a late Greek word "chemia" or "chymeia". The old theory derived this word from the ancient Egyptian "km.t.", that is "black", which would make it the "Black Art" called after the black soil of Egypt, where chemistry was said to have been born.

Still if we analyse such data closer they soon prove to be very deceptive. Thus is it true that the Egyptians called their country km.t, that is "the black land". By this term they tried to express the contrast between the black arable soil of the Nile-valley and the red barren desert sand (dšr.t). But even the Coptic "keme" is never connected with "the black art" or alchemy in any text. Hence Diels' derivation of the word alchemy from the Greek "chyma" that is "casting" seems more plausible. Plutarch (ca. 80 A.D.) mentions "chêmia" once.

Zosimos refers to a *Book of Chemes or Chemeu* (300 A.D.) and later Syriac versions of Zosimos' writings refer to it as "Khumu" or "Khumia". This suggests some connection with the mysterious dwarfs of ancient metallurgy, the "chnumu" of Egyptian tradition. And indeed "chemịa" is now usually derived from "chyma" that is "the casting" of an ingot of metal, from the verb "cheo", to pour or to cast. Hence the terms "chymeutes" (metalworkers) and "chymeutikos" (metallurgical) used by Zosimos. There seems some connection between the shift in meaning and that in spelling, for when Suidas writes "chêmia" in the tenth century he no longer refers to metallurgical art but his term embodies specifically the artificial preparation of silver and gold.

Lately Hermann has argued very plausibly that the "Book Chemeu" of Zosimos might be the very old "Book kmj.t" often quoted by Egyptian scribes from 2000 B.C. onwards and evidently mistaken by Zosimos for the "Satire of the Trades" in which the lot of the smith and other craftsmen is discussed and compared with that of the scribe.

Etymologically there would be no difficulty in deriving "chemeu" from "kmj.t". Zosimos and later alchemists have linked "chemeu" with "chyma" by using the word "chymeutes" (e.g. founders of metal) for chemists. Of late Mahdihassan has proposed a Sino-Indian origin of the word, but this seems a very early penetration of Greek alchemy by Oriental influences which no doubt affected it later.

The influence of preclassical chemical technology

Alchemy, and therefore chemistry too, was the last enduring field of science to emerge from that seething melting pot we call Hellenism. This clash of the civilisation of the Greeks and that of the Ancient Near East contains all the elements that were to build up alchemy, beliefs and notions belonging to:

1. The philosophy and technology of the Ancient Near East;
2. The philosophy and the science of the Greeks;
3. The philosophical tenets of the Iranian and Indian civilisations.

If we first turn to the ancient Near East, it is obvious that the ancient empires of Egypt and Mesopotamia were politically on the down-grade at the period of Alexander's conquests. Persia was definitely in the ascendency. Still the spade has uncovered sufficient evidence to prove how wide and detailed the technological knowledge of the craftsmen of the Ancient Near East already was and remained. The classical world had but few important discoveries to add to metallurgy, dyeing, the manufacture of glass and glazes and similar fields. The detailed study of such crafts reveals many points which are later quoted as characteristic of alchemy.

It is often little realised that the ancient empires of the Near East arose from an "Urban Revolution" which does not yield in importance to the "Industrial Revolution" of the eighteenth century A.D. (1).

This Urban Revolution brought plough agriculture and irrigation, the wheel, the ship, metallurgy of gold, silver, copper and bronze, textiles, and writing, that is the basic crafts of modern technology. The later classical world had very little to add to these early achievements already in common use by 3000 B.C. The story of this urban civilisation which could support the craftsmen with the surplus harvests of the peasants demonstrates a growing importance and specialisation of these craftsmen (2).

The texts of ancient Egypt recovered and published upto now contain very little technological material, though here and there a fragment sheds some light on this gradual growth of skill and knowledge. On the other hand our material form Mesopotamia is much more abundant and the texts from Sumerian upto Neo-Babylonian times illustrate the accumulation of knowledge on minerals, natural products, animals and plants. Both in Egypt and in Mesopotamia the accumulated experience of the craftsmen and their guilds was incorporated in the body of knowledge which in both countries is often cast in the form of onomastica, that is in lists of objects thought to be related and classified according to external characteristics, some of which we would call chemical tests now (3).

The peculiar structure of the Sumerian language invites a nomenclature which adds to the root (the name of the supposed species) a suffix giving certain characteristics to denote the individual member of the species. These suffixes deal with the outer form (IGI = eyestone; NUNUZ = egg-stone; TAG.GAZ = cut stone), colour (GIN = blue; GUG, ḪUŠ = red; SI_7 = yellow), hardness (AŠ = hard), effervescence by acid (ZA.TU), or type of application. The "fire test" seems to have been applied widely and sublimates were known and recovered. This very efficient Sumerian nomenclature led to a classification, which was retained by the later Accadians (Assyrians and Babylonians). An example of the possibilities of this nomenclature as applied to the different forms of bitumen then known is given in Table VIII.

TABLE VIII

Sumerian nomenclature of bituminous substances

ESIR	general for bitumen, crude oil, petroleum
ESIR.LAḪ	"white" crude lake-asphalt
ESIR.IGI	"shining" ("eye") bitumen, asphaltite
ESIR.ḪURSAG	"mountain", rock-asphalt
ESIR.UD.DA	"dry" refined bitumen
ESIR.É.A.	"house" bitumen, mastic

This very clear characterisation of minerals and natural materials has enabled R. Campbell Thompson (4) and others to identify many of them.

This codification of technical experimental knowledge, however, is not the whole picture of ancient technology. Thus a vast amount of experimental data were collected in the ages before alchemy. Familiar household, metallurgical and other terms came to be used indiscriminately to denote certain chemical and technological operations.

Indeed ,most of these terms like those for the apparatus have originally been borrowed from the ktchen and cooking. Like in later alchemical texts we find terms such as "cooking" (bašlu), "leaching" or "washing" (misû) and "roasting" (kalû) being applied to similar operations in different crafts. Nor does this similarity with later alchemical texts stop there.

We have several texts that prove the incidental use of a secret language. Thus a recipe for the manufacture of glass dating back to the XVIIth century B.C. uses "erû (UD-ḪU): eagle" for "êrû: copper". In other texts crude sulphur is called "kibrit ilu nari: bank of the river". Often a kind of craft-jargon is used, abbreviations occur or Sumerian values are used for the Accadian terms not only in technological texts, but also in those of the astronomers and physicians. In medical texts such cryptograms as "lion fat, human fat" for "opium and "blood of a black snake" for "castor oil" are quite common.

We also have many records of efforts to produce synthetic products. Recipes for the manufacture of synthetic lapis lazuli, the "uknu" or blue copper frit so dear to the Assyrians are frequent. Then we also find tablets giving recipes for synthetic copper (K 6246 + 8157 Rs 17; K. 4290 + 9492 + 9477 Rs ff.) and synthetic silver K. 7942 + 8167, 16,22). Meissner (5) believes that recipes for synthetic gold are still hidden amongst the vast mass of unpublished cuneiform tablets.

The archaeologists have found many artificial alloys and Meissner's supposition is not al all fantastic. We must not forget that the art of assaying gold and such native alloys as electrum by the "fining pot" (cupellation) was known as early as 1500 B.C. The touchstone is certainly in use by 600 B.C. Thus the ancients possessed a vast lore of properties of metals and alloys, methods to refine and test them and they synthezised some of them as is proved both by texts and by archaeological finds.

The reasons for this manufacture of alloys apart for their technical merits can be partly found in the aesthetic pleasures which they seem to have derived from the proper combinations of coloured alloys in their works of art. We have recovered examples of inlaywork, metal work of all kinds using natural and synthetic alloys combining their colours artistically. Thus the colours of the gold applied by the Egyptians ranges from bright yellow, grey, various shades of red, reddish-brown, brick colour to dull-purple plum colour and a peculiar rose-pink. Most of these colours are fortitious and due to natural admixtures of varying quantities of silver, copper and iron or to oxidation layers of these baser metals. However, in certain cases this staining is caused by organic matter. The peculiar rose-pink proved to be a heat-resisting translucent coat of oxide of iron. It proved to have been formed by dipping the object in an iron salt solution and heating it (6).

This process was in use many centuries before the oldest written recipes for tinting metals. Old Accadian texts describe methods for staining minerals and stones by cooking them in solutions or embedded in chemicals to obtain fake gems.

The importance of colour to the ancients is manifest from the ancient syllabaries. Thus the 16 Accadian terms for gold embrace no less than nine that refer to a peculiar colour or shade (Table IX). This colour was not only important for the artistic effect to which it was put, it had magical meaning as well. Here we touch a most important aspect of ancient technology. In Antiquity scientific and experimental knowledge is never collected as a body of data from which conclusions are drawn like from the modern body of chemical data. Religion, philosophy and science were still one. Then every bit of chemical knowledge meant deeper knowledge of the Cosmos as a whole, another knot of the "Net of the World", another secret helping to understand the Order of Creation and may be to master Nature (3). The name like any word could mean power. Hence the element of secrecy and initiation gradually creeping into the body of texts. Warnings such as "Let him that knoweth show him that knoweth, but he that knoweth shall not show him that knoweth not!" are found in VIIth century medical texts, but they occur on tablets on glass technology of the Kassite period too, that is over ten centuries earlier. Also there is that insistence on correct copying which springs from the magical potency allotted to the written word. Any inaccurate writing-out was considered heinous.

Many other elements of later alchemy can be found in these early texts. The ancient onomastica and recipes recognize "male" and "female" forms of the same mineral or chemical. The "male" form is usually the harder or darker modification or it is characterized by some peculiar "male" structure. This curious way of ascribing sex to the inorganic world is typical of pre-classical philosophy. Early metallurgy is perhaps the best example of this philosophy. Metals like the earth from which they spring were believed to be subject to the cosmic laws of birth, growth and death. Death and ressurection were their fate and the smith working these "stones charged with mana" performed a rite full of secret dangers. As he conjured the metal out of its ore with the help of the fire-god, his patron, he interfered with the harmonious growth

TABLE IX

The Accadian nomenclature of gold

Accadian	*Sumerian*		*Notes*
1 ḫurâṣu	GUŠKIN (KÙ.GI)	"the yellow", gold	
2 ḫurâṣu ruššu	GUŠKIN-ḪUŠ.A	red gold	
3 ḫurâṣu arḳu	GUŠKIN-SIG$_7$	yellow gold	arḳu is the yellow-green of young shoots (Gr. chloros)
4 ḫurâṣu šadi		mountain gold	
5	GUŠKIN.ŠAR.DA	alloyed gold	
6 ṣarirû	AN.TA.ŠUR.RA	"the ruddy", red gold	from root ṣrr
7 pašallu		gold leaf	compare Syr. p'sêlê = texture
8 šaššu		"the sun-metal"	compare Syr. šamša = sun
9 zâzu		mint-gold	compare Syr. zḳz = coin
10 ṣaidu		refined red gold	from root ṣwd = shine red
11 misû		washed gold	from root msh
12 anaku		(lit. "tin") gold coin	compare Engl. "tin" = gold coin
13 dalbu, daiâlu		"the circulator"	from root djl = walk to and fro?
14 zalḫu		(Subaraean) "the ruddy"	compare Syr. zalîḫê = patina
15 sâmu		"the red"	often combined with 2
16 sakîru		reef gold	compare Syr. s'Kar = rubrum facit

of the metals in the womb of their mother, the Earth. Perhaps the sacrifice of an embryo when building a furnace is an expiatory offering. By giving one life for another the smith avoided the revenge of the Earth-goddess. Or should we see in this rite a means of "charging" the metal with the budding life of the embryo? We have similar texts relating of the offering of sacrifice to embryos by Assyrian glassmakers in the hope that these incomplete beings might assist him during his experiments. It is clear that this belief in the growth of metals is basic for the later alchemical doctrine of the natural perfection of the base metals which gradually turns them into gold. Like the ancient smith the alchemist merely hastens a natural process.

An old Sumerian hymn compares the change of light into darkness with that of gypsum (gaṣṣu, IM.PAR) into bitumen (iddû, ESIR) (Langdon, *Sum. Lit.*, 339) and words like "salmu" (night) and "pisû" (day) are often used for black and white. We also find metals being connected with gods or stars more particularly in Neo-Babylonian (Chaldaean) times. Silver is called the metal of Marduk, gold that of EN.ME.ŠAR.RA, and copper that of Ea (Langdon, PBS, Vol. X, No. 4, 337). A tablet from the British Museum (C.T. 24.49. 3b, K 4349) allots silver, gold, copper and tin to Anu, Enlil, Ea and Nin-a-mal. Again we possess tablets connecting gods, metals, plants and stars (VAT 9874 : K. 11151). The correlation between gods and stars is already quite common and thus gradually the "universal sympathy" principle of the later alchemists grows.

Hence the ancient technologists and craftsmen had an extensive knowledge of ores and metallurgy, the preservation of food and the preparation of fermented drinks, cosmetics and perfumes, glass and glazes, pharmaceutical and medical preparations, colours and dyes. Their lore already contains disconnected elements which are later characteristic of alchemy, being by then moulded into a consistent theory of the structure and changes of matter.

Greek and Iranian contributions to alchemy

On this well-experienced technology tinged with phrases and symbols of the magical world, in which it was born, the clear-cut rational mind of the Greeks impinged even before the tidal wave of their civilisation covered the Near East in the wake of the conquests of Alexander. There is no doubt that certain principles of alchemy were latent in Greek philosophy, even in its loftiest exponents, Plato and Aristotle.

Alchemy recognizes a primary matter, a substratum without qualities, onto which such qualities could be transplanted and thus create the diversity of materials. This theory could be derived very easily from Anaximander's apeiron or from the ekmageion (plastic matter) described by Plato (*Timaios*, 50 C-E) which he himself compares to the odourless substratum of the balsam-cookers (unguentarii) or the potter's clay. Aristotle's materia prima is nor strictly speaking a real substance, but in the hands of the Stoic philosophers it readily obtained the more materialistic character of the archè of the pre-Socratic philosophers. Again the Stoics slowly transformed the Aristotelean qualities too into something materialistic that could be added to or substracted from the materia prima. Thus one of the basic alchemical tenets lay ready for formulation.

The transmutation theory could have been derived from Greek philosophy and its theory of the four elements that build up the world alone, but the germs lay ready in the Near East too. Then there was the trend to think in pairs of contraries so dear to Greek philosophers, which may have been borrowed quite early by Pythagoras from the East, but which was later absorbed by Greek thought as was the theory of cosmic sympathy which bound the macrocosm to the microcosm. The strong bonds between microcosm and macrocosm, the "so above, so below" plays a large part in alchemy.

Finally we must remember that though we always speak of the mathematical and logical structure of Greek philosophy as contrasted with that of the Near East, there was a strong animism at the bottom of Greek thought if we except the atomists like Democritus and Epicurus. Even Aristotle had to people his macrocosm with immaterial forces which closely ressembled the Platonic mechanism of heaven. This animation of the physical world belonged to the earliest strata of Greek thought and it came to the fore again as the Stoic philosophers started to blend thought with the magical world of the Ancient Near East.

A third stream of thought originated in Iran and was propagated by the Magi and later the Mazdians, the followers of Zoroaster. Its strong dualism presented this world as a clash between good and evil, between the world of the spirit and that of the flesh, and left man to choose his side. Purification and redemption were possible with the helps of the priests. In its wake came the belief in the possible apotheosis of man and the coming of a redeemer who would liberate the world from its throes. The tenets of the Magi showed a tendency to syncretism or absorption of foreign faiths and elements as they pushed westwards.

In order to assess the effect of Iranian (and may be Indian) philosophy on the Near Eastern body of knowledge we may pause for one moment to summarize the present views on the origin of astrology, a science so closely related to alchemy that we can hardly point out an alchemist who is not a believer in astrology at the same time. This is due to a very large overlapping of the basic tenets on which astrology and alchemy are built. Astrologers prophesying from celestial phenomena either the fate of the land or that of the king (wars, harvests, etc.) were common enough in early times, their art belongs to the class of mantics like hepatoscopy and other forms of omina. But the new horoscopical astrology was based on the mathematical calculation of celestial conjuctions and positions. It also required a standardized zodiac of twelve signs of equal length and came to birth in the sixth and fifth century B.C. Only then were Mesopotamian mathematics and astronomy ready for this task. Side by side with the older omen-astrology it continued its course until through the centuries the horoscope-astrology prophesying personal fate from the positions of the planets and stars at birth or conception won out (8).

This new astrology, the birth of which we can establish from dated cuneiform tablets containing horoscopes, is based in the following philosophical tenets:

a) The harmony between macrocosm and microcosm;
b) The descent of the soul through a star to the body and its return to heaven through the same star, a belief of Avestan origin;
c) The worship of zodiacal signs.
d) Solar theology, as contrasted with the older Moon theology of Mesopotamia;
e) Number mystics, later so dear to the Pythagoreans and Gnostics.

The astrologers believed that through the study of numbers and of the Universe man could become divine and immortal. Initiation quickly becomes typical of these "mathematicians" as they are called in later Antiquity.

We need not go into details of early horoscopical astrology but we can see from their doctrine how closely related the atmosphere in which astrology grew up was to that which we find in alchemy. Would it be too bold to believe that alchemy too, like astrology had its origin in that seething period of the sixth and fifth century B.C. when the basic elements of Hellenism arise to flower when Alexander's conquest and pacification of the Near East has created the political situation so favourable to the universal spread of these beliefs?

The incubation period of alchemy

This "incubation period" of astrology, which we believe to be that of alchemy too, is characterized by the penetration of the Greeks into Mesopotamian world, politically dominated by the Persians. Cnidian physicians serve at the court of the Persian King of Kings. Thus Ktesias, who was doctor to Artaxerxes II, met Plato about 367, before the latter left for Sicily for the second time. Greek architects and scientists travelled and worked in the Near East and brought back new data and strange beliefs to Greece (9). In truth the period of Persian domination of the Chaldeans and the rest of the Near East started the intimate contact and exchange of ideas between Greece, Iran and Babylonia. Accadian words are absorbed into the Greek language along with the material objects such as plants which they denote. This body of knowledge was still growing and hardly systematised as it will be later on by generations of Greeks at the universities of the Hellenistic world. Through these Greek authors the vague and fluctuating theories of these pre-Hellenistic centuries find their systematic treatment in the earliest Hellenistic books on astrology and alchemy. Plato has formulated this so well in his *Epinomis'* "What the Greeks may have taken from the Barbarians they have always carried to higher perfection."

At the same time the Babylonian world even before the fall of Babylon (538 B.C.) is penetrated from the East. The Magi, the exponents of Zoroaster and the Avesta, bring to Mesopotamia the Iranian dualistic philosophy of the struggle between the great powers of Good and Evil in which man has to choose his side and fight. The Magi were the intellectual leaders, who carried their dualism, their theory of the harmony between macrocosm and microcosm and their belief in the innate powers of man to reach perfection by gnosis, for beyond the pales of the Persian Empire .Though they are often dubbed Chaldaeans in later Hellenistic literature their great influence is clear from such writings as the "peri Magon" written by Hermippos of Alexandria about 200 B.C. From many sources we know that the Magi had penetrated as far as Lydia by the fourth century. By then their influence in Babylonia was so firmly established and mingled with the more ancient Accadian doctrines that it is impossible to disentangle them. Not only Greeks like Kallisthenes bring back scientific data and theories from Mesopotamia but Chaldaeans reach Greece and teach there. Socrates' death was predicted by a Syrian sage and Euripides' fate cast by a Chaldaean. Pythagoras is said to have been taught by "Zaratas the Chaldaean" in which the Semitized form of the name of Zoroaster is typical for the interpenetration of culture then going on in Mesopotamia.

The old scientific centres of Mesopotamia, Babylon (and Borsippa), Assur, Kalah, Niniveh and Uruk absorbed many Persian beliefs during Neo-Babylonian times. This amalgamation went on through Seleucid time upto the beginning of our era (10).

Though generally writing their scientific documents in imperfect Sumerian and Accadian, these scientists and priests spoke Aramaic and even Greek, another proof of the syncretic character of this period (11).

Its results crystallised in the Greek writings of the early Hellenistic period. Then the great Hellenistic centers like Pergamom, Antiochia and above all Alexandria tended to attract the leading personalities and schools. This may be the explanation why the earliest alchemical and astrological documents (many of which we know by name only) originated in Alexandria. We have definite proof that astrology (and probably alchemy too) did originate in Mesopotamia and not in Egypt, which country was its traditional home, but where the direct links with the past are lacking.

Just as astrology was shaped in this period it seems that alchemy as the transmutation of natural bodies, more especially metals, first took shape in these regions, where the Armenian mountains housed a very old metallurgy which already believed in the natural growth and evolution of metals from "base" lead to "perfect" gold.

In the troubled Seleucid era some of the five Babylonian centres somehow managed to survive, but they declined and were extinguished in the Parthian wars. Their scientific activities were partly continued in Western Mesopotamia where centres like Nisibin, Edessa and Harrān (Carrhae) flourished right up to the Arab period. The Sābi'ans of Ḥarrān were branded as "star worshippers" and "Chaldaeans" by the young Arab civilisation and their activities were only ended by the Mongols in the thirteenth century A.D.

The codification period of alchemy

The interchange of ideas started in Chaldaean times was certainly accelerated when Alexander the Great conquering the Persian Empire began to weld the Greek world and the Near East into a new unit.

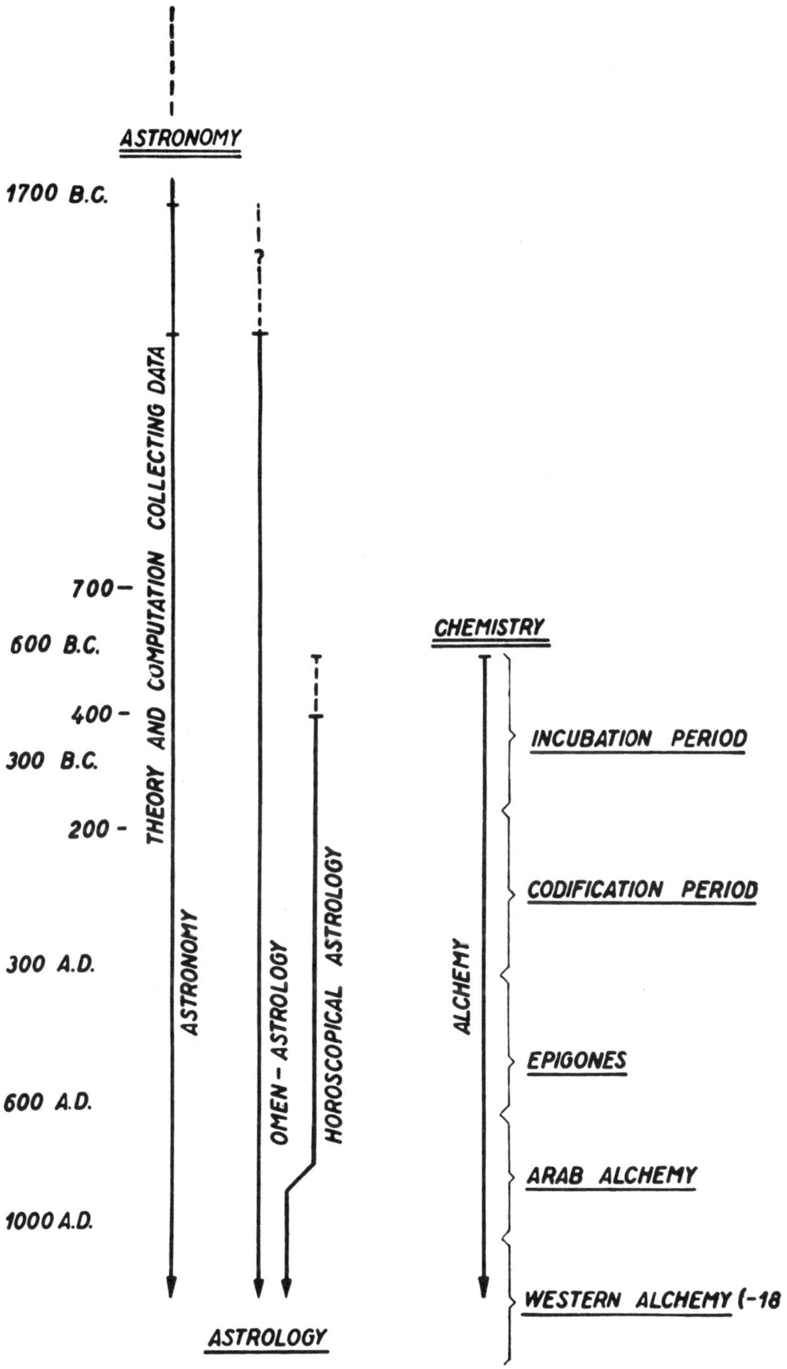

Fig. 31.
The birth of alchemy.

Even if he failed to attain his end because of his untimely death, there is no doubt that Hellenism was a melting-pot of cultures. Iranian, Chaldaean and Egyptian tenets and beliefs were grafted onto Platonic and other Greek philosophies to form Neo-Pythagorism, Neo-Platonism and like schools, repeopling the world with the old gods and powers, knowledge of whom could be obtained by initiation and study or revelation. These schools showed a strong tendency towards syncretism and in the first centuries of our era had absorbed so many elements from each other that they can hardly be distinguished. Gnosis with its strong amalgamating tendencies and its elasticity of belief was the serious enemy of the young emerging Christian faith. Its blend of quasi-rational and mystical doctrines tended to offer an easy explanation and mastery of the problems of the world.

It was logical that these activities would gravitate towards the great centres of Hellenistic civilisation such as Alexandria, Antiochia and Pergamom. Of these Alexandria was far the most important, being the great centre of the trade between the Mediterranean, India and the Far East. Whilst the other Hellenistic Empires were frequently disturbed by wars between the successors of Alexander, Egypt generally enjoyed long stretches of peace so important to the development of science, which was moreover furthered by the foundation and liberal endowment of the Academy by the Ptolemies. The institution comparable to the Royal Society or the Académie des Sciences was the centre of Greek science during Hellenism and naturally Alexandria attracted other less academic branches of science. If therefore we find the earliest records of alchemy in Alexandria, this does not prove their traditional Egyptian origin, but simply that they found there in an international city the proper conditions for peaceful development so stimulating to science.

Therefore after its incubation period in Mesopotamia, alchemy with other theories came to Alexandria, the great pooling centre of the Hellenistic sciences, where it was crystallised into a clear-cut doctrine by the logically-minded Greeks. This *codification period* starts with the *Physika* of Bolos Democritos of Mendes (200 B.C.) (12), a Neo-Pythagorean. Wellmann has pointed out that we still find the traces of this original work in the *Codex Marcianus* and he has demonstrated that Bolos' treatise was composed of four books dealing with the making of gold, the making of silver, the making of gems and the manufacture of purple. Bolos takes his data and philosophy from Egyptian, Jewish, Babylonian and Persian sources. Thus Ostanes the Magus becomes the teacher of Democritos in the art of alchemy. This is quite in line with the traditional stories about Ostanes whom king Xerxes sends to Egypt to teach the Egyptian priests alchemy. His methods are clearly Babylonian, he changes "bodies" by embedding them in chemicals which are then to penetrate the body by prolonged heating of the mass (cementation). The traditional Egyptian method seems to have been the "projection" (by sublimation, etc.) of chemicals onto the body to be changed, which may first be conditioned by "roasting" (13). Stapleton has recently argued that the *Treatise of Agathodaimon* may be the oldest alchemical document we possess. It seems to have been written in Seleucid times (4th-1st cy. B.C.) in Northern Mesopotamia, may be in Harrān, the old city of the Moon-god Sin and lays down that the Stone should only be prepared from mineral substances, showing great knowledge of metallurgy and geology. This attractive hypothesis should be further investigated (14).

Bolos also uses the accumulated experimental data of the ancient technologists, who seem to have compiled textbooks on alloying metals, faking precious metals and gems, dyeing (Baphika) and similar technological operations. It is still widely believed that the Papyrus Leidensis X and the Papyrus Holmiensis are such technological handbooks, though in a rather mutilated form. However both Pfister (15) and Reinking (16), whose excellent book was unfortunately never carried beyond the proof-stage, have proved that these two papyri can never have represented real textbooks or collections of recipes for dyers and metallurgists, as in this form the recipes can not yield any practical results. We actually have a Coptic mss. dealing with the practical dyeing of textiles which dates from the VIIth or VIIIth century A.D. (Berl. Pap. 8316) but which has a format very similar to the Papyrus Holmiensis. Hence it was possible for Reinking to reconstruct the original dyeing recipes which are reproduced in a mutilated form in the Leiden and Stockholm papyri.

It may be true that there were original Egyptian documents dealing with the manufacture of mock jewelry and precious stones. Still it should be remembered that before Alexandria manufactured such cheap jewelry, Syria and Phoenicia had specialized already for centuries in their manufacture from alloys and glass and had also produced substitutes for the expensive Tyrian purple. The two papyri were certainly written in Egypt, for instance the word "argyros" (silver) is given as "asemon" (Egyptian *d*3m for elektron). But they are no longer technica manuals on colouring metals and textiles. Already Bolos and the all chemists, keenly interested as they were in the changes of matter, had studied the colouring of metals and textiles and had used the information from practical dyers of textiles and their terminology to describe the tinging of metals and stones. Terms originally denoting the degreasing, mordanting (stypsis) and dyeing (baphé) of textiles are now applied to the apparent change of form of metals. Thus they call the transmutation into silver baphé and that into gold katabaphé (superdyeing). Other terms like varnishing, bronzing and waxing are derived from ancient recipes for colouring metals. They therefore made extracts from the technological handbooks and the recipes of the craftsmen of their period and produced series of experiments which might allow them to study and understand this transmutation. In these alchemical treatises of which the two papyri form part many of the practical recipes are reproduced only partly. New terms or cryptograms have been introduced to denote the perfectly innocuous chemicals of the dyers and metallurgists. Their aim is no longer a practical one but the study of the transmutation of matter, a philosophical one. The colour change, which to these alchemists seems the proof of this transmutation, is effected by dyeing, by varnishing and alloying and quantatively controlled in some cases by the increase of weight.

Colour had from the beginning been one of the most obvious means for the craftsmen to identify ores, metals and other materials. They identified the material by and with the colour, for the colour did not only indicate their nature. Colour had a magical meaning, not only were properties ascribed to metals and stones because of the colour, but certain colours were prescribed for magical figures and effects. Coloured circles play a part in the mantics of the "very insignificant sect called the Ophites" (Origen). The Anglo-Saxon *Leech Book of Balt and Cild* (X) and later books on magic simply continue the classical traditions of Alexander of Tralles and Pliny and prepare the way for such books as the *Sworn Book* of Honorius.

In the earliest alchemical texts the colour must have had both a magical and a practical meaning. For the colour did not only indicate the appearance but also the inner nature of the metal or compound. Transmutation was achieved in different steps. The first step began with an "earth", with some unidentifiable solid or an alloy or base metal such as lead, "tetrasomy" (lead, tin, copper and iron) or "metal of magnesia". On to this "body" the qualities of liquidity ("water") or fusibility, brilliancy ("air") and "fire" had to be imposed. Hence the original metal or material had often to be broken down to a "body", a degeneration which was often accomplished by fusing with sulphur and which was called blackening or *melanosis*.

Then followed a whitening or *leukosis*, which often emant a fusion with the "ferment" or "seed of silver", in which such ingredients like mercury and arsenic played a part. It was the counterpart of the original faking of silver and its alloys (argyropy). The third step consisted of yellowing or *xanthosis*, the counterpart of the original faking of gold (chrysopy) in which the "water of sulphur" often played a part.

The final stage was the production of a violet or purple colour, the *iosis*, in which the violet ferment would change the gold through and through into an "ios of gold" which was the permanent tincture which when cast upon common gold would produce more.

So intense is the preoccupation of the alchemist with the transmutation of metals that he seldom pauses to ponder on the nature of the colours. Galen had already credited the Greek philosopher Democritus with the view that such qualities as colours are sensed by us from the concourse of the atoms, but that they do not reside in the atoms themselves. Such physical theories hardly ever occur in alchemical tracts. The early alchemical manuscripts are therefore devoted to the study of matter and its changes and they stand apart from the true technological handbooks or collections of recipes, for they are not of a practical but of a philosophical nature.

We should never forget that the seemingly practical recipes of the Leiden and Stockholm papyri did not stand alone. These two papyri were found in a grave at Thebes (Egypt) together with the magical papyris nos. XII and XIII (published by Preisendanz) and the magical papyrus Leidensis V, which contains symbolical names for plants and stones. Bolos' book contained both the philosophical tenets and the tests of the alchemists. The Leiden and Stockholm papyri seem to go back to a Baphika written by Anaxilaos of Larissa in Egypt, where he had been banished by the emperor Augustus about 2 B.C. The practical (Isis, Iamblichos, Moses, Ostanes) and philosophical (Maria the Jewess, Comarius, Hermes, Cleopatra) treatises, part of which were ascribed to mythical persons such as Toth, were composed in the early centuries of our era. The codification period ends with Zosimos of Panopolis who in the third and fourth century summarizes the entire alchemical doctrine and literature. Already in his writings we find a strong religious factor, there we read that salvation can be obtained by the Great Work.

The theory of alchemy contained many elements which predisposed is for absorption into philosophical or theological speculation. This is clear even from a rapid summary of its main tenets.

In Aristotle's philosophy a substance (ousia) was built up of matter (hyle) and form (morphè), which latter contained the essential individual qualities. In the hands of the Hellenistic alchemists, influenced by Neo-Pythagorean, Neo-Platonic, Stoic, Jewish and Chaldaean tenets all qualities and sensations, which help us to characterize chemical compounds, were turned into subtle types of matter. Heat and cold were thought to be the presence or absence of a "matter of heat". Thus the qualities were turned into extremely thin fluids (pneuma or spiritus, literally breath), vapors which penetrated the materia prima and changed its properties. Therefore attention was concentrated on change as a phenomenon rather than on details of individual changes through chemical reactions.

This strongly counteracted the rise of a mechanistic view of chemical change in which the phenomena are connected with the form, size and arrangement of primary particles. Also the concentration on the phenomenon of change led to comparison with changes observed in man and the use of such terms as generation, birth and death for chemical change. Dualistic philosophy introduced the theory of countraries such as body and soul, matter and form, matter and energy, passive and active, male and female. The union of form and matter is thought to create the chemical compounds we know. Hence the later picture of the generation of such compounds as the marriage of Sun and Moon or King and Queen. The final creation of the Philosopher's Stone is called the "Magnum Opus", but also the "Chemical Wedding". Therefore alchemy is strongly qualitative.

The theory that substances could be reduced to a sufficiently simple matter to be given the form of any other substance was basic for alchemy. For thus the alchemists was sure that any substance was capable of being changed into any other. Hence the first step is stripping a substance of its form and recovering the materia prima. This "ousia" is often lead or a "tetrasomy" that is an alloy of iron, copper, tin and lead. This dead body (soma or corpus) needs a life-giving breath or pneuma to give it the desired form and turn it into the final product. This pneuma is described as anything from a gas or vapor to the Holy Ghost, generally it should be volatile and change the colour the of soma.

The generation of the bodies in nature was thought to be due to preformation and purposeful creation. Hence the generation of new chemical substances is a creation too and the pneuma is compared with sperma, the breath of life, the breath of heaven and fermentation. It is also called a tincture as it gives the essential colour.

Early recipes purported to have been invented in Mesopotamia tend to achieve this by embedding the primary substance in the reagents and heating the whole to make the pneuma penetrate the soma. The Alexandrian chemists seem to make the pneuma penetrate the soma. The Alexandrian chemists seem to have preferred "projection", that is applying the pneuma by sublimation or condensation onto the soma. Such operations were gradually more and more executed under the influence of suitable planetary influences as calculated with the help of astrological hours and seasons.

Alchemy therefore rested on three basic assumptions: The possibility of transforming any kind of matter into any other, the need of "corrupting" a substance into primary matter before the transformation could be achieved and finally the power of the subtle not wholly immaterial pneuma to evoke and generate new forms, which perfection already lay in the nature of these forms.

The very doctrines and operations of the alchemist lent themselves most willingly to symbolical interpretation by the philosopher or the mystic.

The period of the epigones

After Zosimos we enter the *period of the epigones* in which the corpus of alchemical literature as we know it is finally codified and commented upon by a host of authors, who have nothing new to contribute. They are mainly Neo-Platonists or Gnostics to whom alchemy is part of their religico-philosophical doctrine. Original contributions to alchemy begin to flow again when the Arab scientists enter the scene.

Such "alchemists" like Stephanos and his school (8th. cy. A.D.) use the transformation of metals as a symbol for the regenerating force of religion in transforming the human soul. It is doubtful whether such "alchemists" had any practical laboratory experience. Their interest lies in the religious sphere, they mix alchemical texts with prayers, invocations, moralising paragraphs and allegories. The chemical operations are entirely subjected to their allegorical-symbolistic interpretations. Others seem to have used music to accompany alchemical operations and thus achieve the proper harmony between body, soul and the music of the spheres. Therefore by the end of the period of Greek alchemy we have three types of chemists which co-exist and which should be well distinguished. There are still the many craftsmen working in various branches of what we now call chemical technology, then the students of the structure and changes of matter, the alchemists proper, and finally the philosophers and mystics who use alchemical theories and data in their speculations.

Zosimos is still an Egyptian but already the centre of alchemy tends to shift to the north in the third century A.D. The parts of Zosimos' important encyclopaedia of alchemy which survived show us that in his days experiments still played an important part in the Art.

This is much less so in the writings of the later Greek alchemists (after 400 A.D.) which tend to become more and more mystical and devoid of practical experience. Alchemy is then no longer an Egyptian but a Syrian art, part of its writings being in Aramic. This Syrian school commented on the older writers who were believed to have possessed the secret of transmutation and they tried to rediscover it reinterpreting their writings. Still Aeneas Baraeus, a Greek theologian of Gaza, Syria, spoke of the alchemists as a recognized group of artisans with a certain standing and a lore of procedure (484 A.D.).

This experimental attitude was never lost even when Syrian alchemy sank into the depths of mystical speculation. This is to be seen from Byzantine alchemy, which continued the Syrian traditions. In his Chrysopoeia Michael Psellos based his philosophy on pseudo-Democritos and he cited Zosimos and Theophrastus, but he also referred to practical metallurgy and technology. Still his book gives the impression that Psellos never entered an alchemical laboratory himself, and had his knowledge from books only. The experimental side of alchemy was finally revived by the Arabian alchemists.

Syria throughout the centuries remained the centre of alchemy and the focus of cultural activity in general and with it Mesopotamia. Religious conflicts in the Byzantine Empire first led to the expulsion of the Nestorians and then of the Monophysites who came to Syria and Mesopotamia and even settled in the countries to the East. Greek philosophical and scientific works were translated into Syriac, the language which had displaced Aramaic that had ruled the Near East in early Hellenistic.

An important problem of the codification period is the possible contact between Greek and Chinese alchemy along the trade-routes of later Hellenism. Or were there no earlier relations than between Arabic and Chinese alchemy? Mahdihassan has recently discussed this point in various essays.

The exact interrelation between Chinese and Greek or Arabic alchemy is yet far from clear, mainly because far too little original Chinese texts have been translated. The earliest really alchemical data from China go back to the third century B.C. It should, however, be remembered that earlier Chinese philosophy (for instance that embodied in the ritual to which the emperors had to submit in their daily routine) already speculated on the Two Contraries, Yang, the male active, fiery principle and Yin, the negative, earthly and dry principle. Early ritual also correlated the four seasons, points of the compass, elements, colours and tastes. On the other hand we also find a system of Five Elements (Wu-hsing) (water, fire, wood, metal, earth). Though counterfeiting gold by alchemical methods was forbidden as early as 175 B.C. the Chinese alchemists were not after the production of the rare metals but wanted this gold as a substance to obtain longlevity. The gold was to be used for shaping vessels from which potions of longlevity were drunk, or for the manufacture of the "pill of immortality". Here again cinnabar, having the colour of blood, played a large part together with mercury, the "living metal".

Alexander the Great by pushing into Afghanistan had established Hellenism on the outposts of the Near East whence the desert routes led to China and beyond. Influences travelled to and fro along with merchandise. We are not yet able to judge the extent of this intellectual interchange for the lack of Chinese alchemical texts. Certain streaks of Chinese alchemy do appear in Arabic alchemy, one of them being the idea of the "elixir". Still there is no need to believe that Chinese alchemy arose due to influences from Alexandria or vice versa. As far as we can judge now Chinese alchemy arose from autochtonous ideas simultaneously with Greek alchemy, though interchange of ideas was not only possible but obvious from the little evidence we possess now, and may have started early.

Though we are of course far from understanding all aspects of alchemy in its codification period we should now also direct our attention to the incubation period proper. It will be necessary to cooperate with cuneiform scholars but the harvest is most promising. There are still scores of cuneiform tablets yet undeciphered and now indifferently dubbed "medical, pharmaceutical, chemical and technical" which promise to yield important data on the rise of alchemical doctrines codified in the early Hellenistic period by Bolos and the Greek alchemists of Alexandria.

Notes

1. Forbes, R. J., *Man the Maker* (New York, 1950)
2. —, *Professions and Crafts in Ancient Egypt* (*Archives Internationales d'Histoire des Sciences*, No. 12, 1950, 599—618)
   A History of Technology (edit. Ch. Singer c.s.) (Oxford, 1954, Vol. I)
3. —, *Man and Matter in the Ancient Near East* (*Archives Internationales d'Histoire des Sciences*, Vol. I, 1948, 557—573)
4. Campbell Thompson, R. C., *Dictionary of Assyrian chemistry and geology* (Oxford University Press, 1936)
   —, *Dictionary of Assyrian Botany* (British Academy, London, 1949)
5. Meissner, B., *Babylonien und Assyrien* (Heidelberg, 1925, Vol. II, p. 385)
6. Lucas, A., *Ancient Egyptian Materials and Industries* (London, 1962, p. 233)
7. Forbes, R. J., *Metallurgy in Antiquity* (Leiden, 1950, p. 85 ff.)
8. v. d. Waerden, B. L., *The Thirty-Six Stars* (*Journal Near Eastern Studies*, VIII, 1949, 6)
9. Goossens, G., *Artistes et artisans étrangers en Perse sous les Achéménides* (*Nouvelle Clio*, 1949, 1/2, 32—44)
10. Bidez, J., *Les écoles chaldéennes sous Alexandre et les Séleucides* (*Ann. Inst. Phil. Hist. Orient.*, Vol. III, 1935, 41—89)
11. Goossens, G., *Uruk sous les Séleucides* (*Acad. Belgique Bull.* 27, 1941, 222—244)
12. Wellmann, M., *Die Physika des Bolos Demokritos und der Magier Anaxilaos aus Larissa* (*Abh. Preuss. Akad. Wiss.* 1928, Phil. Hist. Kl. No. 7)
13. Bidez, J. et F. Cumont, *Les Mages hellénisés* (Paris, 2 vols, 1938)
14. Stapleton, H. E., *Probable sources of the numbers on which Jabirian alchemy was based* (*Archives Int. Histoire des Sciences*, 1953, No. 22, 44—59)
15. Pfister, R., *Teinture et Alchimie dans l'Orient Hellénistique* (*Seminarium Kondakovium*, Vol. VII, 1935, 1—59)
16. Reinking, H. *Die Färberei der Wolle in Altertum* (Leipzig, 1939)

Bibliography on ancient alchemy

Bidez, J., *Dernières recherches sur l'histoire de l'alchimie en Grèce, à Byzance et en Egypte* (*Byzantion* 13, 1938, 383—388)

Browne, C. A., *The poem of the philosopher Theophrastus upon the the Sacred Art* (*Scient. Monthly*, 1920, 193—214)

Browne, *Rethorical and Religious Aspects of Greek Alchemy* (*Ambix*, Vol. II, 1946, 129—138)

Chikashige, M., *Oriental Alchemy* (Tokyo, 1936)

CROISSANT, JEANNE, *Matère et Changement dans la Physique Ionienne* (*L'Antiq. Class.*, 1945, 61—94)

DARMSTAEDTER, E., Artikel „Chemie" (*Reall. f. Assyr.*, Bd. 2, 88-9)

DAVIS, T. L., *Primitive Science* (*J. chem. Educ.*, 1935, 3—10)

DÜRING, I., *Aristotle's chemical treatise Meteorologica, Book IV* (Göteborg, 1944)

ELIADE, M., *Cosmologie si alchimie babiloniana* (Vremea, 1937)

EISLER, R., *Der Babylonische Ursprung der Alchemie* (*Chem. Z.* 49, 1925, 577, 602)

FESTUGIÈRE, A. J., *La révélation d'Hermes Trismégiste I* (Paris, 1944)

—, *Sur des textes alchémiques* (*Rev. Etud. Grec.*, 1949, 235—236)

FESTUGIÈRE, A. J., *Alchemica* (*L'Antiquité Classique* 8, 1939, 71—95)

FORBES, R. J., *Short History of the Art of Distillation* (Leiden, 1948)

FORBES, R. J., Artikel „Chemie" (*Reallex. f. Antike und Christentum*, 1954, cols. 1061—1073)

FRIEDLÄNDER, P., *Structure and destruction of the atom according to Plato's Timaeus* (*Univ. Calif. Publ. Philos.*, XVI, 1949, 225—248)

GANSCHINIETZ, R., *Hippolytos Capitel gegen die Magier, Refut. Haer. IV.* 28—42 (*Texte Unters. Gesch. Altschr. Lit.*, 39. 2 (1913))

GANZEMÜLLER, W., *Wandlungen in der geschichtlichen Betrachtungen der Alchimie* (*Chymia*, III (1950), 143—155)

GANZEMÜLLER, W., *Die Alchemie im Mittelalter* (Paderborn, 1938)

GIGON, O., *Das Problem der Wissenschaft in der Antike* (*Universitas* I, 1946, 1073—1084)

GOLDSCHMIDT, G., *Der Ursprung der Alchimie* (*Ciba Z.*, V, 1938, 1950—1988)

—, *Ein Beitrag zur Ursprungsgeschichte der Alchimie* (*Cahiers de Frontonex*, 1947, 101—126)

HASCHMI, M. Y., *The beginnings of Arab alchemy* (Ambix, vol. IX, 1961, 155—161)

HERMANN, A., *Das Buch kmj.t und die Chemie* (Z. Aeg. Sprache vol. 79, 1954, 99—105)

HOPKINS, A. J., *Alchemy, Child of Greek Philosophy* (New York, 1934)

HOPKINS, A. J., *Defence of Egyptian Alchemy* (*Isis*, 28, 1938, 424—431)

LAGERCRANTZ, O., *Papyrus graecus Holmiensis* (Uppsala, 1913)

LEVEY, M., *Chemistry and Chemical Technology in Ancient Mesopotamia* (Amsterdam, 1959)

LEVEY, M., *Research sources in ancient Mesopotamian chemistry* (Ambix. vol. VI, 1958, 149—154)

LI CH'IAO-P'ING, *The chemical arts of Old China* (Easton (Pa.) 1948)

LIPPMANN, E. O. VON, *Some remarks on Hermes and Hermetica* (*Ambix*, II, 1938, 21—25)

MAVROMICHAELIS, C., *Alchimie aux premières siecles de l'ère chrétienne* (*Suisse contemporaine*, 1943, 862—870)

MAHDIHASSAN, S., *Alchemy and its connection with Astrology, Pharmacy, Magic and Metallurgy* (Janus vol. 46, 1957, 81—103)

MAHDIHASSAN, S., *Alchemy in the light of its names in Arabic, Sanskrit and Greek* (Janus vol. 49, 1961, 79—100)

METZGER, H., *Alchimie* (*Revue de Synthèse* 1938, 43—53)

PREISENDANZ, K., *Zur Ueberliefrungsschichte der spätantiken Magie-* (*Zbl. Bibl. wesen*, Beiheft 75, 1951, 223—240)

REHM, A., *Zur Ueberliefrung der griechischen Alchimisten* (*Byz. Z.* 39, 1939, 394—434)

PFISTER, R., *Teinture et Alchemie dans l'Orient Hellénistique* (*Seminarium Konda-Kovium* VII, 1935, 1—59)

PLESSNER, M., *The Turba Philosophorum* (*Ambix* vol. VII, 1959, 159—163)

REITZENSTEIN, R., *Zur Geschichte der Alchemie und des Mystizismus* (*Gött. Gelehrt. Anz.*, 1919, 1—27)

SHEPPARD, H., *The redemption theme and Hellenistic alchemy* (*Ambix* vol. VII, 1959, 42—46)

SHERWOOD TAYLOR, F., *The Alchemists* (New York, 1949)

SHERWOOD TAYLOR, F., *Symbols in Greek alchemical writings* (*Ambix* I, 1937, 64—67)

—, *The visions of Zosimos* (*Ambix* I, 1937, 88—92)

—, *The Alchemical Works of Stephanos of Alexandria* (*Ambix* I, 1937, 116—139; II, 1938, 39—49)

STAPLETON, H. E., *The Antiquity of Alchemy* (*Ambix* vol. V, 1953, 1—43)

STAPLETON, H. E., *Ancient and Modern Aspects of Pythagoreism* (Osiris XIII, 1958, 12—53)

WELLESZ, E., *Music in the treatise of Greek Gnostics and Alchemists* (*Ambix* IV, 1951, 145—158)

WILSON, W. J., *Origin and development of Greco-Egyptian alchemy* (*Ciba Symposia* III, 1941, 926—960)

WINDERLICH, R., *Das Zeitalter der Alchemie* (*Angew. Chemie* 60, 1948, No. 10)

ZAHND, H. & D. GILLIS, *Chemical knowledg. in the New Testament* (*J. chem. Educ.*, 23, 1946, 90—97, 128—134)

Chapter III

WATER SUPPLY

Though the story of planned and organised water supply belongs to that of urban life, its earlier stages reach back far beyond the limits of prehistory. Water supply like food is essential for human life, but mankind during many centuries depended on natural sources of water supply only.

Natural Sources

The Upper Palaeolithic and Mesolithic food-gatherers of Europe have been found to camp near springs or on the banks of rivers, which also supplied them with fish (1). The same holds true for the ancient Near East.

The earliest farmers congregated round springs (or "tites"), usually funnel- or saucer-shaped holes at the bottom of shallow, oval or circular depressions. These springs were usually enclosed in wooden casings in prehistoric Europe, seldom more than 5 m deep and no more than 2 m wide. Short sections of hollow tree-trunks are sometimes used as casings. The enclosed spring of mineral water at St. Moritz is one of the oldest in Europe (2) and many more mineral springs have been used for centuries. Rainwater was not generally collected in Europe before the Iron Age, it seems.

In the ancient Near East the collection of the scarce rainwater and of spring-water also goes back to very early times. Nearly all the Old Testament hill-site cities depended on springs at the foot of the town-mound. Thus En-Rogel (The Dragon Well) and the Spring of Gihon (Spring of the Steps) watered David's City (3). Many of the wells in the ancient river-valleys and in the Arabian Desert were not man-made but natural springs. Indeed desert travel largely depended on such springs supplemented by hand-dug wells or other water supplies (4). We have the curious story told by Herodotus (5), that the empty wine jars were collected in Egypt and gathered at Memphis:

"They are filled with water by the Memphians, who then convey them to the desert track between Egypt and Syria" where they were placed at certain regular intervals. This custom still persists on the old slave-trading route between the Sudan and Egypt and on old routes between Dakhla and 'Uweinat (in the Western Desert). This ancient attempt to supplement natural resources brings us to an older one, the well.

Wells

Unfortunately most authors do not properly distinguish between the natural "spring" and the man-made "well". The word "well" should be confined man's attempts "to obtain water from the earth, vertically below the spot where it is required, when it is not obviously present at the surface" (6). Many of the so called "holy wells" are in reality enclosed springs, often deepened much later (7).

Wells are typical for urban life and irrigation farming, when greater congregations looked for larger supplies of water. There was a common opinion that the earth floated on an "Ocean", which figures in ancient religious literature as the "waters of chaos" from which the earth and civilised life sprang and also as the abode of the dead and the powers "below". As such water had acquired a special ritualistic and symbolistic aspect in Antiquity and even in later centuries (8). At the same time close observation of nature and experience in mining and tunnelling had accumulated some practical knowledge of water-finding. Real geological knowledge was not available. Such strictly scientific principles were applied to the choice of the site of a well for the first time in Derbyshire in 1795 (6).

Primitive peoples still possess an uncanny instinct of finding water. The Roman author Vitruvius gives us a few more scientific means of spotting a good supply (9). Looking for water vapours rising close to the ground, inspecting the type of soil and observing the vegetation should be followed by trials, in which bronze or lead vessels or unburnt clay pots are buried for a short time and inspected for water.

These ancient wells were never drilled but hand-dug. The majority was circular and all were steined, that is they were lined with stone, brick or wood. The wells of Harappa and other Indus valley cities were found in private houses (diameter 50—75 cm) as well as in market-squares (upto 250 cm in diameter). In Mesopotamia (Ur, Niniveh, Mari, etc.) wells were dug about the same time e.g. Eannatum's well at Lagash (Tello), or slightly later. In Egypt wells were first used to supplement natural irrigation, then they are dug in cities too. Some of these wells attained considerable dimensions, such as the big well of Hermopolis which 65′ wide for about 50′ and then narrowed down to 32′ for another 65′. Then there is the big Ptolemaic well at Matruh and the later well of Cairo which was no less than 280′ deep. From 1500 B.C. onwards the ancient Egyptians learned to drive horizontal passages (or adits) into the strata at the bottom of wells to increase their output. It is clear that such efforts sprang from a long experience in mining. The well of Lachish (Tell Duweir, Palestine) was 250′ deep.

In most cases, however, these wells were some 0.50—2.00 m wide and less than 30 m deep. Sometimes pottery rings take the place of brick linings. This is particularly true of the wells of Crete and their Mycenaean counterparts on the mainland of Greece (Philakopi). The Acropolis of Athens had two steined 60′ wells and the Romans preferred brick or concrete lined wells, but in military camps or temporary posts helped themselves with old wooden barrels, plankings between four solid corner-posts or morticed plank constructions (10). These wooden casings were adopted in the rest of Europe between 200 B.C. and 100 A.D. The Etruscans usually produced bell-shaped wells with a constricted mouth, the Romans preferred an even batter or a gradually narrowing section.

In desert regions the well was essential for life and carefully protected though available to friend and foe. Thus Uzziah "built towers in the desert and digged many wells, for he had much cattle" (11). Sometimes the spring or well was enclosed in a little building through the lattice-work of which one could laddle out the water, like the "seqaya" of the Arabian desert.

Water was drawn from these wells by hand, as is still the custom in most cases nowadays. Lengths of bamboo, gourds, shell dippers, pottery jars and buckets have been in use for centuries. In deep wells relays of men and women down to the water's surface passed the water-jar on to each other. With the evolution of water-raising machinery new methods were introduced. Ox- or camel-drawn wells appear early in history. When Sennacherib records his building of "the palace without rival" and his care for the water supply of Niniveh he writes (694 B.C.): "that daily there might be abundant flow of water of the buckets I had bronze cables and pails made and in the place of the mudbrick pillars I set up great posts and cross-beams over the wells" (12). "The rivers of Babylonia... were spread over the cornlands by hand or by the help of swinging beams (shâdufs or swipes)." (13) As irrigation techniques improved we find this machinery used for water supply in towns, by Greeks and Romans too. Thus the water for the thermae (baths) of Stabiae (Pompeii) comes from a well (section 2 × 3 m, 25 m deep) which stands besides a basin (1.5 × 2.0 m) in a little building which also contains a long narrow room with a treadmill. This treadmill worked by two slaves could be made to work two sets of chains of bailers (3.5 l each) hanging in the well. The output would be about 2—3.6 m^3/hour (14).

Storage of water, cisterns

Water was not only taken from springs, wells and rivers. In the ancient Near East rainwater was soon collected as purity was not only found to be important for drinking, but also for such industrial purposes as dyeing, fulling and washing flax and linen. The other types of water were used for bathing, flushing drains and similar purposes.

Primitive and prehistoric peoples stored water in water-holes. The "dewponds" on the chalk hill tops of Sussex ond Dorset may belong to these primitive cisterns for rainwater, though some believe them to be of Roman date (15). In prehistoric Europe timber-lined shafts were sunk into strata of impermeable clay to store rainwater cisterns (16). The ancient Near East, however, built masonry cisterns or sank them into the rocky subsoil often giving them the shape of a bottle with a narrow opening to prevent pollution of its contents. "And they hewed them out cisterns" (17) and Rabshakeh, the messenger of the King of Assyria promises Israel "drink ye everyone the waters of his own cistern" (18). Some of these are of considerable size such as the "Pools of Solomon" at Jerusalem, the royal cisterns which are some 13 m deep. The largest one is 190 m × 69 m × 16 m deep. These caves often pillared and reached by a flight of stairs as common in Palestine and Syria were imitated by the Greeks and Romans.

Ancient Cnossos (Crete) had its cisterns like Mari (Mesopotamia) and Mycenae (19). Apart from these rock-hewn cisterns the Greeks and Romans built cisterns of masonry or concrete with tun-vaults and pillars. Smaller clarifying basins were often attached to such cisterns In Hellenistic and Imperial Roman times the large cities like Alexandria and Byzantium built huge cisterns. In the latter city largest measures 141 by 73 m and has 420 columns. These huge cisterns built by Valens and Justinian still serve the city of Constantinople (20).

Conduits and pipes, sewers

Even the Romans preferred to take their watersupply from wells, but when water had to be brought from a distance irrigation and drainage technique had taught that one should preferably rely on gravity flow. The open duct or conduit, which could be cut into rocky soil, may have been fairly costly, but it was easily cleaned and covering slabs could protect the water from pollution. In prehistoric Europe, again, timber was cheap and plank channels or hollow trunks formed the basic form of conduits. Thus the water from the St. Moritz spring was transported in two lines of hollow tree-trunks with a diameter of 1.10—1.40 m and 0.80—1.05 m respectively (21).

Open terracotta conduits were first used on the mainland of Greece by the Mycenean builders of Tiryns and Philacopi, though the Cretans of Cnossos had preferred closed ducts to prevent the accumulation of silt (22). The famous Roman aqueducts are none but open ducts with covers carried down from the hills and sometimes carried by tiers of arches.

However, in very early historic periods we already find preference for earthenware or stone pipes used for the transport of smaller quantities of water and for drains. Earthenware pipes, some of them flanged, are in favour in the houses, drains and sewers of such Indian cities as Chandu Daro and Mohenjo Daro. The earthenware pipes and masonry sewers, water closets and drains of some Mesopotamian towns such as Mari are still in perfect working order. Already these pipes are made in standard length. Thus the pottery rings for drains and sewers in the Indus cities and those in Mesopotamia are 30 cm high and 11 cm in diameter (23).

The palace at Cnossos contained very sophisticated water pipes, 76 cm long, 2 cm thick and slightly tapering towards one end (diameter 17 cm—8.5 cm). These pipes were flanged on the widest end and joined in such a way that the loops on these pipes were used to strengthen the joint by external cording. Also the flow of the water was direct from the thick to the slim end of the pipe.

Metal pipes come early too in history. A length of 400 m of copper pipes was found in the mortuary temple of Sahurê (Abusir, Egypt). The pipes, each some 40 cm long, were made from strips of 1.4 mm hammered copper. The ends of these 47 mm pipes were joined without soldering, probably by hammering on a wooden core. This line was fitted in the hollows cut into the solid flagstones. Bronze lines were also used to convey water from the mainland to the island of Tyre and at Motye on Sicily. The Greeks used earthenware, stone, bronze or lead pipes. Lead pipes seldom exceeded a diameter of 5 cm. With proper sealing earthenware pipes were proved to hold upto 50 atm. pressure, though Roman engineers preferred to encase them in concrete. Earthenware bends were used in Greece too (Olynthus; fourth cy. B.C. onwards). Stone lines (10—30 cm diameter) were much in favour with the Greek cities of Asia Minor, they could stand pressures of 4—15 atm. and were sometimes used in syphons. The oldest stone syphon was found at Patara (Lycia). The older form of earthenware pipe was fluted like those of Cnossos, but the classical Greeks preferred cylindrical pipes, slightly tapered at one end for fitting into the next one.

Though the Roman engineers used lead pipes for pressure lines such as syphons they preferred wooden or earthenware ones for supplies in private houses. Bronze pipes were too expensive and used extensively only in the villas of the rich of the Roman Empire at Rome, Pompeii, Baiae, etc., who sometimes even indulged in silver pipes. The Roman wooden pipes were used with iron collars to strengthen the joints, and patching was done with lead plates. In the works on the Alban Lake (396 B.C.) stone collared pipes were used (90—95 cm length, diameter 36 cm, overlap 6 cm). Though lead pipes were fairly common in the city of Rome, the engineers were well aware of the danger of lead poisoning (24) and tried to avoid them.

The lead pipes of the Roman Water Board should be mentioned as the first series of industrial products to be standardized. In discussing the water-supply of Rome we shall have occasion to point out that these series were based on the "quinaria", a lead pipe made by bending a lead strip with a width of five Roman digits round and soldering the ends (25) with lead-tin alloys (26). Pliny summarizes the Roman view on pipes in these words: "The most convenient method of making a water course from the spring is by employing earthen pipes, two fingers in thickness, inserted in one another at the point of junction. The one that has the higher inclination fitting into the lower one- and coated with quicklime macerated in oil. ...The proper length for each leaden (pressure) pipe is ten feet, and if the pipe is five fingers in circumference its weight should be 60 pounds; if 8 fingers one hundred; if ten, 120; and so on in the same proportion.

A pipe is called a "ten-finger" pipe when the sheet of metal is ten fingers in breadth before it is rolled up; a sheet one-half that breadth giving a "five-finger pipe" (27).

During the Middle Ages similar materials were used. Only when metallurgy could produce cast iron by increased air-supply to the smelting furnace with a more efficient use of heat, a new material was available. By the middle of the fifteenth century were cast-iron water-pipes, produced in double-valve moulds, used in Germany (Dillenburg Castle, 1455) (28).

The ṣinnōr of ancient Palestine

The cities in the river valleys of Egypt and Mesopotamia could easily be supplied with water from the river or local wells in times of war. This was very much more difficult in the case of the ancient cities of Palestine and Syria which were built on rocky hill-tops which had their water-supply from springs at the foot of the mound. Hence the engineering skill required in irrigation and mining was there turned to tunnelling from the midst of the city's fortifications to the spring.

From this strategic need arose the typical "ṣinnōr" or watertunnel which originally was a flight of stairs leading down a shaft to a tunnel which was a secret approach to the spring outside the city. Later attempts were made to enclose and defend the spring and to carry its waters to the foot of the shaft by means of a conduit located in the floor of the tunnel. Thus the chance of the enemy taking the city through the tunnel, as when David took Jerusalem from the Jebusites (29), was considerably lessened. Such hidden and protected wells or springs have also been found in other parts of the world, for it was essential to protect the water supply of a city.

Most of the water-tunnels of Palestine can be dated only approximately, moreover they underwent many changes in the course of their history, but they all seem to have been started in the Bronze Age (before 1200 B.C.). The ṣinnōr of Gezer seems to be the oldest (30). Its stairs lead down to a 130′ passage which at the spring in a 80′ × 28′ cave about 95′ below the city level (130′ below the present one!). The Canaanites, who controlled the city's water supply were only subdued by Egyptian armies, whence it passed to the hands of Solomon as part of the dowry of Pharaoh's daughter.

The Megiddo water system was studied and published in detail (31). The city-shaft with stairs led to the 150′ tunnel which led to the well.

For here the spring had gradually been deepened into a well situated in a cave 5—7 m high, part of which housed a guard. A skeleton of such a guard was found on the spot. At a later date the water from the well was led by a conduit to the foot of the shaft at the immediate disposal of the city population.

The actual details of the great shaft at Lachish (32) have not yet been described in sufficient details, but we know much more about the "conduit of the upper pool in the highway of the fuller's field" (33) and how "Hezekiah (727—669 B.C.) made a pool and a conduit and brought water into the city" (34). Before his days the Spring Gihon had been impounded in a reservoir used to water the King's Gardens south of Jerusalem. The tunnel may have arisen out of the threat of invasion by the Assyrian King Sennacherib in 705 B.C. From the exposed Spring Gihon in the Kidron valley this tunnel conducted its waters to the more protected junction of the Tyropoeon and Kidron valleys and spills them into the Pool of Siloam (53′ × 18′, and 19′ deep). The tunnelling was hastily and badly done for a 350 m distance was covered by a tortuously winding 533 m tunnel, dug as usual from both ends (35).

Other less examined shafts which may belong to this type of water tunnel, exist at Gibeon (el-Jib) and Ibleam (Khirbet bel'Ameh) (36). Though these Bronze Age ṣinnōrs are good engineering feats showing skill in mining and levelling techniques, they were not meant to carry large quantities of water over really big distances such as the aqueducts which Sennacherib built inspired by his neighbours of the Armenian mountains.

The qanat and the birth of the aqueduct

For many centuries to come lack of geological knowledge prevented the ancients to locate subterranean sources of water supply. The first faint hint of some elementary geological knowledge comes to us from the story by Procopius on the well dug in the fortress of Baras (Mesopotamia). As it would have been impossible to include a spring in the nearby foothills in the fortifications the engineers dug a well in the fort and expected to find water when they had reached the same level as the spring, which they actually did. The Assyrian engineers had none such knowledge (or should we call it a hunch?), however they learnt one way of tapping subterranean water supplies from their northern neighbours. The planning and execution of large projects show the existence of practical applied geometry and triangulation techniques, which deserve high praise if we take into account the means at their disposal (37), which can only have been the most primitive sighting devices as used by their astronomers.

During his eighth campaign King Sargon II of Assyria invaded Urartu (the present Armenia) and devastated the complex irrigation system around the town of Ulḫu (near Lake Urmia) (38). Notwithstanding this action Sargon greatly admired the efforts of the King of Ulḫu which he describes in these words: "Following his ingenious inspiration (lit. his heart's desire) Ursâ, their king and lord... revealed the water-outlets. He dug a main duct which carried flowing waters... waters of abundance he caused to flow like the Euphrates. Countless ditches he led out from its interior... and he irrigated the fields." The report on its thorough destruction also specifically mentions the blocking up of the duct or canal. Close study of this text (39) confirms an earlier suspicion (40) that the qanāt, the tunnel tapping water from the

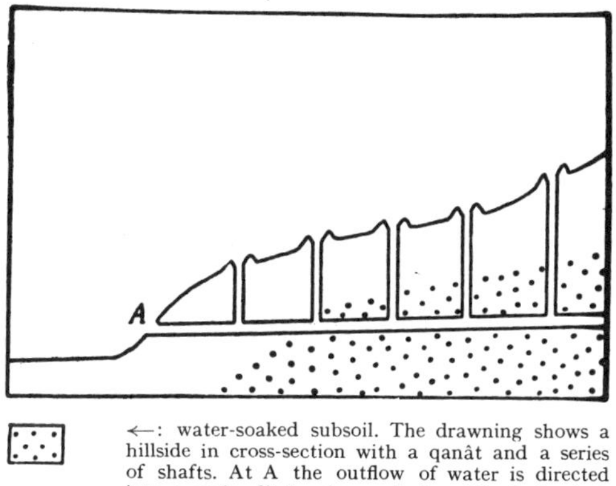

←: water-soaked subsoil. The drawning shows a hillside in cross-section with a qanât and a series of shafts. At A the outflow of water is directed into a main ditch whence it is utilized for irrigation purposes.

Fig. 32.
The qanât or water-tunnel.

foothills to transport it to distant fields by gravity flow, originated in ancient Armenia.

The qanāt (I)*, already widely spread in Antiquity, is essentially the same as the adit driven horizontally into the mountain side by the ancient miners. With its vertical air-shafts (and inspection shafts) at regular distances it seems derived from very old mining techniques. The old flint-mines of the New Stone Age were a set of vertical shafts which when reaching the stratum containing the flint-nodules required, are often connected by horizontal galleries and their branches. This technique was often applied by later miners too. Also the miner would be the best man to spot his main enemy, underground water supplies and water bearing formations. Hence the ancient miner would possess both the necessary practical geological knowledge and the skill of

* The Roman figures refer to the notes on page 188.

constructing horizontal adits, in the hillside to tap its water. Armenia is one of the oldest mining and metallurgical centres in the Near East and thus may well claim the invention of the qanat.

These qanats were already widely used in Persia in the period of Achaemenian Kings (41). Megasthenes found them so frequent in northern India about 300 B.C., that there were "officials which inspect the closed canals from which the water is distributed into conduits, in order that all may have an equal use of it" (42). These adits, penetrating the hillside for several hundred feet until they reach the water-bearing gravels, deliver their water to conduits which may be 4 to 60 km long. The adits (average 2′ × 4′, slope 1—3%) have vertical air-shafts with a diameter of 0.60—1.00 m which may upto 120 m high (43). The qanat has now spread to Trans-Caucasia, Afghanistan, Kashgar and Chinese Turkestan, their construction is the privilege of century-old guilds of specialists (II).

They also penetrated to Arabia, may be as early as Herodotus' time, for he reported (44): "The Arabian king, they say, made a pipe of the skin of oxen and other beasts reaching from this river Corys all the way to the desert, and so he brought water to certain cisterns which he had had dug in the desert to receive it. It is a twelve days' journey from the river to this desert track. And the water was brought through three different pipes to three separate places."

The Persians certainly introduced this new technique to Egypt, and more particularly to Kharga Oasis (45). Though king Sheshonk (945—924) already mentioned a "chief of irrigation", the Persians opened up the subterranean water-supply by adits and rendered the oasis habitable. The qanats were probably contructed during the reign of Darius I in conjunction with the erection of the Hibis temple in the same oasis. The mines and agriculture were further developed by the Ptolemies and the Romans who organised the Riff trade. It spread into the Sahara, even to southern Morocco, and was used at Roman Garama (Tripolitana). The Arabs even imported their "fogarra" into Sicily, where the fountain-men of Palermo still use terms of Latin, Byzantine and Arabic origin (See also vol. II of these Studies).

The classical water-tunnels, even if they were not true qanats, often have very similar constructions. The Greek ducts or the tunnel in the aqueduct of Samos have their air- or inspection-shafts and Vitruv says: "If there are hills between the city and the fountainhead... tunnels are to be dug... Air shafts are to be at distance of one actus (120′) apart." (46).

Sargon may have destroyed these qanats of Urartu (III), but he brought the idea back to Assyria where it was used before the Persians

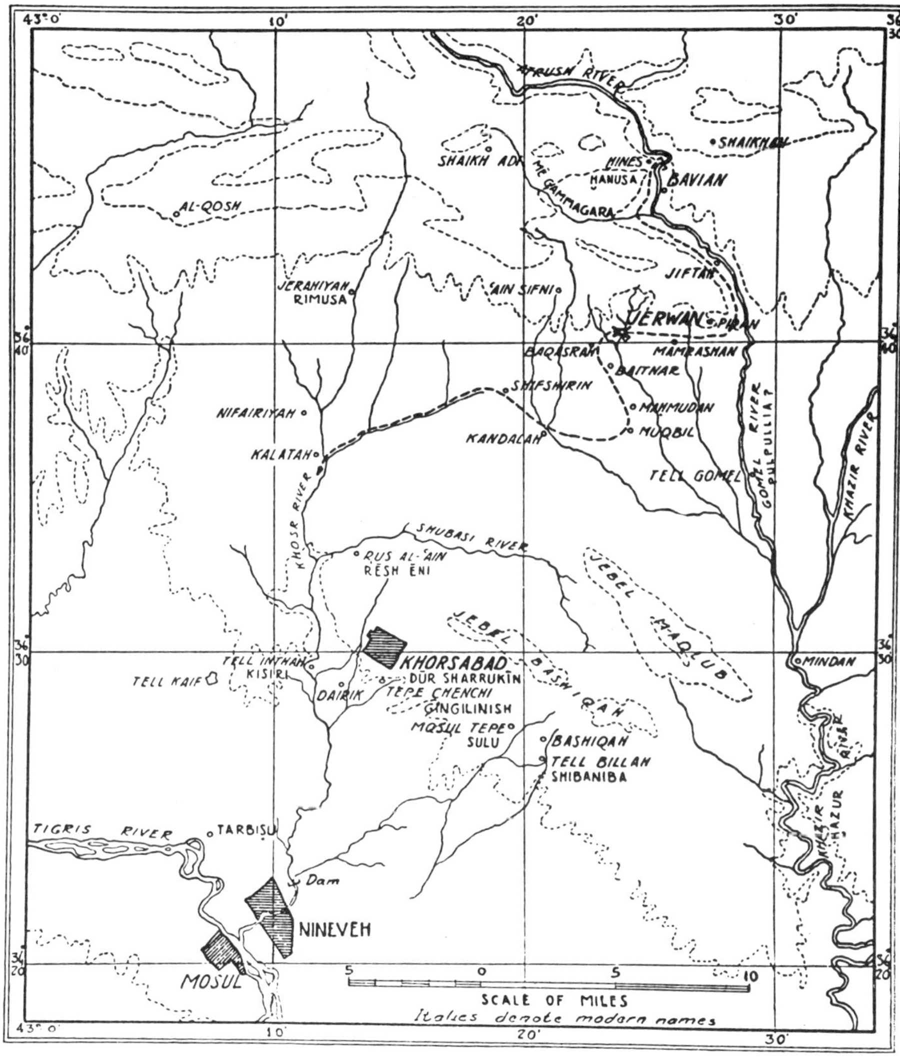

Fig. 33.
Map of the region north and east of Niniveh, showing the approximate course of Sennacherib's Bavian-Koshr Canal.
(Photo Jacobsen and Lloyd, *Jerwan aqueduct*).

were famous for it. Sennacherib applied it in building supplies of Niniveh and Arbela. Nebuchadnezzar II used qanat-like constructions for the basis of the ziggurat of Borsippa (47) and of a wall east of Babylon, and such a drain was also found at Ur.

The first large water-supply system which Sargon's son, Sennacherib, built was to provide water for Niniveh and the royal palace of Chorsabad (48) (Dur Sharrukin), and to irrigate the fields around the old and the new capital. We read from his annals and inscriptions that this work proceeded in three stages.

First Sennacherib started (703 B.C.) to throw a weir or dam across the Khosr river near Kisri, north of Niniveh. "The river Khosr, whose waters from of old took a low level and none among the Kings my fathers had dammed them as they poured into the Tigris;— to increase the productiveness of the low-lying fields, from the border of the city of Kesiri (IV), through the high and low grounds I dug with iron pickaxes, I ran a canal; those waters I brought across the plain (around) Niniveh and made them flow through the orchards in irrigation ditches." Weirs as mentioned by Sennacherib were nothing novel. The lake of Homs (2 × 6 miles) on the Orontes is nothing but the partly-silted up reservoir behind the dam (20′ high, 500′ long) built in the river by Ramses II (c. 1300 B.C.) which raised its level some 16′—17′. We also know the huge dam at Marib (S. Arabia) which probably dates from 1000 B.C. and many more from Hellenistic and Roman times. This 10-miles canal provided the plains west of the Khosr with water.

In 694 B.C. "to explore the waters which are at the foot of Mt. Musri (the modern Jebel Bashiqah, north-east of Niniveh) I took the road and climbed up and with great difficulty came to the city of Eluminakinne. At the head of the cities of... I saw pools and enlarged their narrow sources and turned them into a reservoir. To give these waters a course through the steep mountains, I cut through the difficult places with pickaxes and directed their outflow on to the plain of Niniveh. I strengthened their channels, heaping up (their banks) mountain-high, and secured those waters into them. As something extra I added them to the Khosr's waters for ever. I had all the orchards watered in the hot season, in winter a thousand fields of alluvium, about and below the city, I had them water every year. To arrest the flow of these waters I made a swamp and set out a cane-brake within it." This entailed the canalisation of some 18 water-courses mentioned in further texts and also the building of two dams (of square stone blocks) across the Khosr defile near Niniveh itself, behind which the swamp held the waters (49).

The third and most ambitious scheme dates of 690 B.C. "The bulk of those waters, however, I led out from the midst of Mt. Tas, a difficult mountain on the border of Urartu (Armenia)... Now I at the command of Assur the great lord, my lord, added unto it the waters of the mountains on its sides from the right and left and the waters of... (three) towns in its neighbourhood; with stones I lined the canal and Sennacherib's Channel I called its name." Actually Sennacherib canalised the Atrush river (the upper course of the Gomel River, a tributary of the Khazir river) upto Bavian where this new canal met a slightly older system which tapped water from the Atrush and conducted it to the older Khosr river system.

A weir had been placed obliquely across the stream to dam it and to allow superfluous water to pass over it. The gorge of Bavian thus formed a natural reservoir, from the south-western corner of which the outlet ran into the canal which then follows the course of the foothills, winding slowly down with a 1 : 80 slope until it joins the Khosr river near Kalatah. Thus the canalised Atrush finally sent its waters through the older system to Niniveh at a distance of over 55 km. Unfortunately the number of workmen employed on it is lost but the inscriptions tell us: "In one year and three months I finished its construction."

Sennacherib tells us that he prepared for the opening-ceremony and sent down priests and presents for the gods, when the "gate of the river like a... was forced open inward and let in the waters of abundance. By the work of the engineers (V) its gate had not been opened when the gods caused the waters to dig a hole therein." It has been supposed that this "gate of the river" was a sluice situated in the canal when as a long artificial cleft, 7 m wide, cut in the cliffs on the rights bank of the Gomel it has to pierce a protuding rock to join the older canal leading the waters away down South. It is more probably that the engineers after tunnelling this rock had blocked it again temporarily for the opening ceremony, a measure which failed through the force of the rising waters. We have no indication whatever that the Assyrians knew sluices of the hinged door or any other type. It seems more probable that the "bâb nâri" was simply the "mouth of the canal." Holding up the waters of such gravity-flow canals by means of hinged gates or slide-doors would not stem the flow of the water down the canal. They could have been used to draw water into branch-canals. Also the "lock" would have involved the problem of opening and closing such gates or doors against the pressure of the onflowing waters, which could not be solved without winches or such heavy machinery, no traces of which have been reported (49). Hinged wooden doors are said to be used for this purpose in Ceylon in the second century B.C. (21). Probably to the surprise of the trembling engineers Sennacherib took this mishap as a propitious sign that "the gods prospered the work of my hands" and loaded them

Fig. 34.
View of the Jerwan aqueduct (1933) during excavation.
(After Jacobsen and Lloyd).

with "linen and brightly coloured woollen garments. Golden rings, daggers of gold I put upon them."

The canal from Bavian to the Khosr river crossed several small brooks and valleys by means of arched aqueducts such as that at Jerwan which was studied in detail (50). This 300 m aqueduct consists of a slightly concave duct, 12 m wide, enclosed by two walls 1.60 m high and 2.50 m thick. The duct consisting of three layers of limestone blocks is carried by a dyke or arches, resting on a foundation of rough boulders, built up from limestone cubes (50 cm cubic) and mortar of burnt limestone, capped by a 40 cm stratum of limestone chips mixed with river sand and burnt limestone (ratio 1 : 1.5 : 2—4). On the total length of 300 m there are 14 buttresses to strengthen the dyke and its aqueduct.

Sennacherib built another water supply for the town of Arbela (Erbil) (51). He joined "three rivers in the Khâni Mountains to waters of the springs on the right and left sides of the river." These three tributaries of the Bastara river meet in the Darband pass, where some 200 blocks of masonry represent the remains of a weir near Qala Mortka, some 20 km north of Erbil. The water raised by this weir is tapped by a qanat, a series of vertical wells linked together at their base by an underground tunnel. The opening of the tunnel (120 × 112 cm) is lined with neat ashlar masonry for about 600 m when it gradually widens (to 270 cm) though the lining is now only 50 cm high. At Erbil, some 150′ below Qala Mortka, the tunnel changes into an open duct. At the entrance of the tunnel it is pierced with holes which may have accomodated ties to hold a sliding door to regulate the flow of water.

These Assyrian qanats and aqueducts have been described at some length as they anticipate the earliest Greek aqueducts by some century, though their main purpose was providing water for irrigation (52).

THE AQUEDUCTS OF GREECE AND THE SIPHON

Long pipelines had also been built before the Greeks. Ancient Sidon obtained its water from the source of the river Zahrany near the southern end of the Lebanon. In a duct of earthenware pipes (encased in lead?) the water was carried down gorges and ravines on a winding 15 mile course to the town of Sidon proper.

The Mycenaeans were the first to build water-supply systems in ancient Greece and the Greeks have built aqueducts, tunnels and pipelines since to convey good drinking water to their important towns. Their earliest medical works already voice the general opinion that good water is essential for the health of the population (53) and bathing was certainly popular in ancient Greece.

Prominent amongst these is Hippocrates' *On Airs, Waters and Places* which displays shrewd observations on quality of water and general health. The Greeks apply freely all the systems used in the ancient Near East. Thus Athens' aqueduct bringing water from Mount Pentelicus is of the qanat-type, its underground channel has a vertical airshaft about every 50 feet.

A second type very common in ancient Greece was the pipeline supported by stones and boldly tracing the shorted way from its source through valleys and tunnels. Its oldest example may be the aqueduct of Samos on which Herodotus (54) reports: "The Samians are the makers of the three greatest works to be seen in any Greek land. First of these is the double-mouthed channel pierced for an 150 fathoms through the base of a high hill; the whole channel is seven furlongs long, 8′ high and 8′ wide; and throughout the whole of its length there runs another channel 20 cubits deep and 3′ wide, where through the water coming from an abundant spring (near the village of Agirdes) is carried by its pipes to the city of Samos. The designer of this work was Eupalinus son of Naustrophus, a Megarian."

The actual tunnel is 1100 m long (55), but though started from both ends hardly deviates from a straight line. The Greek geometers had solved these practical problems far better than Hezekiah's engineers at Jerusalem. Heron of Alexandria tells us how such careful triangulation can be achieved with the help of a simple dioptra, which is nothing but a water-horizon mounted with sights on a circle divided into 360°. He also solves the problem of excavating the vertical shafts ending in the tunnel (56).

In the highly accidented regions of Asia Minor and other parts of the Greek world the Greeks introduced a daring new solution, the siphon. It avoided the expensive tunnelling or the long tortuous windings of a gravity-flow duct. Siphons had been used for many centuries to mix liquids such as wines and even as drinking tubes. Their principle was still unsolved. The first attempt to explain the siphon was made by Heron (57) though he based it on the false hypothesis that "Nature abhors a vacuum." The fact that a liquid attained the same level in communicating vessels was also well-known. However, using the siphon as a shortcut across a deep valley entailed high pressures in the lower part of the siphon.

The aqueduct of Pergamom built during the reign of Eumenes II is a typical example of the use of such a siphon (58). The watersupply is conducted from a spring in the mountains (+ 1174 m) to two settling tanks at Hagios Georgios (+ 360 m) east of the city by gravity flow. From these tanks however, a pressure line brought the water in two siphons through two valleys (+ 177 m and + 195 m) over a ridge (+ 235 m) to the citadel (+ 332 m). This means that pressures upto 20 atm. had to be overcome in the two siphons.

It is not known how the Greeks got rid of possible air-pockets in these siphons, nor what material these pipes were made of. In the city earthenware pipes were used for the distribution of the water from the basin in the citadel, but they may date from Roman times. The pipes forming the siphon ran through the 1′ holes in perforated stone slabs (4—5′ long, 24″—27″ wide, 8″—10″ thick) standing on their edges about every 4 feet. Bronze or wooden pipes have been suggested,

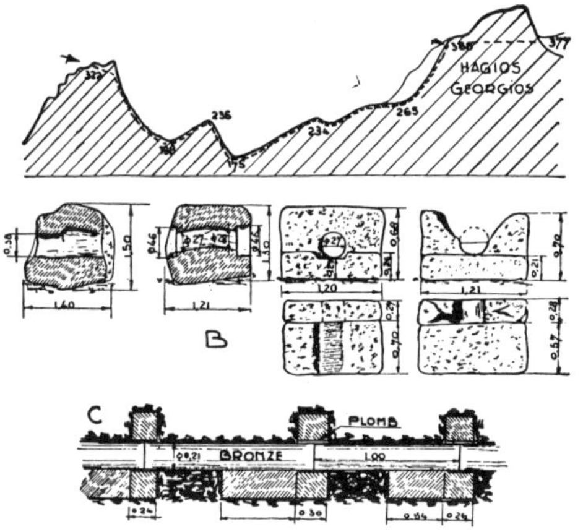

Fig. 35.
The Pergamom watersystem and its details.

lead may have been used as in Roman times or even earthenware pipes, but unfortunately none of the ancient materials have ever been tested against such pressures as 20 atm. As the pipes have all disappeared metal was probably used and then taken down when the siphon was replaced in Roman times by an aqueduct bridging the valleys on tiers of arches but no longer supplying the highest parts of the city with water.

Siphons and bold tunnelling are typical for the water-supply systems built by the Greek tyrants and the Hellenistic kings from Sicily and southern Italy to Asia Minor. In Roman times the siphon was used only in some few cases, probably because of leakages and the relatively poor materials available for high pressures.

The Roman aqueducts

The Roman aqueducts are widely known because of the graceful arches with which they span the valleys of Spanish, French and Italian rivers and the plains of the Roman Campagna. One is apt to forget that the greater part of their ducts is underground, as will be seen from the chronological survey appended to this essay. They are not only discussed by classical authors but they have been studied in great detail by modern archaeologists (59). Their water was meant for household use, for the baths, fountains and public conveniences, for flushing sewers and other hygienic purposes (60).

Water was seldom used for irrigation in the classical world, for the dominant agricultural system was "dry farming", a method of conservation of the spring rain until summer by suitable working of the soil. Irrigation was applied only if after all the climate was too arid or if the type of crop demanded a great quantity of water. Thus in Numidia and Mauretania (Algeria) springs were used to irrigate the hill-country, streams were dammed, reservoirs, ponds and underground cisterns were connected with a system of aqueducts and canals to supplement dry farming. Numerous decisions of the lawyers on water rights form a testimony to its importance in certain regions of the Empire. An inscription with the plan of an aquaduct (the Aqua Crabra near Tusculum, Italy) gives the names of the properties, the number of pipes supplied and the hours they could be opened (61). In the Near East all kinds of water-lifting machinery replaced the aqueducts.

In certain exceptional cases aqueducts were built for industrial purposes such as hushing at the gold mines "arugiae" in Roman Spain (62). Water-wheels were sometimes moved by water from special aqueducts such as the Aqua Traiana at Rome or the one built for the military flour-factory of Arles (63).

The story of the Roman aqueducts begins with the building of the Aqua Appia by the blind censor Appius Claudius. This underground aqueduct was built at state expense for the purpose of bringing pure water into the thickly settled quarters of the city (64). The next one, the Anio Vetus, was built from the spoils taken from king Pyrrhus, in the same way the Aqua Marcia is said to have been financed from the booty taken at Corinth and Carthage (65). Early Republican times saw the extension of Rome's water supply as well as some local aqueducts like that of Pisaurum (66). By 125 B.C. the water supply had been doubled following the rapid expansion of Rome. In the same period the town of Aletrium built an expensive water-system which included a siphon of earthenware pipes embedded in blocks of concrete (67).

In the troubled days of the later Republic (first century B.C.) the aqueducts were neglected and Augustus had to repair the channels and ducts which had fallen into decay and to build new ones (68). The Imperial period was a great era for building and several emperors excelled in enlarging the water-supply of their main cities. Apart from Augustus there was Claudius whose huge aqueduct took 14 years to

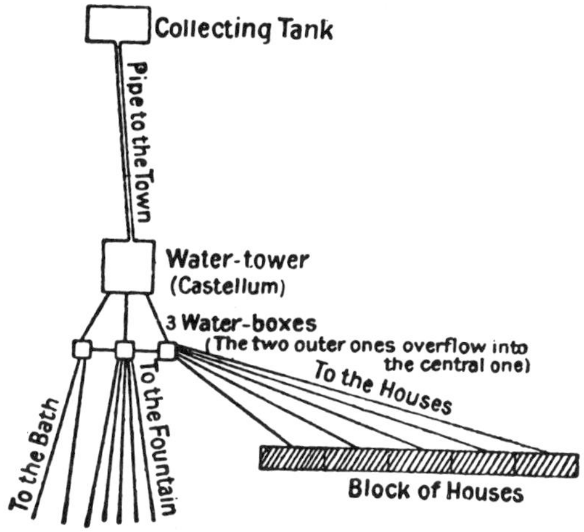

Fig. 36.
The Roman system of distributing water.

build. His engineers had to grapple with the serious problem of quarrying blocks of tufa along the Anio river and hauling them several miles to the building site. No less than 40.000 waggonloads were transported every year (69). Hadrian built a large number of aqueducts all over the Empire (70) and the later emperors continued this policy upto the days of Justinian (71).

The provinces profited by this policy. In Africa the whole extent of the aqueducts, canals and wells is not yet completely known (72). In Egypt Augustus built an aqueduct for Alexandria (73), which water was purified (74). In the rest of the country, however, the ancient irrigation machines were used for water-supplies of towns too. Thus the military camp of Babylon (near Memphis) to which "the water is conducted up from the river by wheels and screws and 150 prisoners are employed in the work" (75). In 113 A.D. the town of Arsinoë built a water-wheel which raised water into reservoirs which fed the townfountains, a bath, a brewery and two Jewish synagogues (76).

In Asia Minor, the old Greek initiative was reawakened by the Roman Empire (77), Sardis, Ephesos, Pergamon, Smyrna, Miletus and Nysa got new aqueducts. In Spain the large aqueducts of Merida and Segovia are well-known, the bridge of the latter is over 1000 m long and 50 m high. In Gaul the aqueducts of Nimes and Arles were famous, Lyons had several including the siphon of Beaumont which carries the water 17 m over a river with 7 lead pipelines (diameter 270 mm, 35 mm lead) for which some 10.000 tons of lead were used (78). In Germany the aqueducts of Cologne, Bonn, Mainz and Trier were studied (79), that of Bonn comes from the Urfvalley, 127 km away.

In Britain the Roman aqueducts of Lincoln, Dorchester and other towns have disappeared. That of Worcester supplied private houses, through separate sluices for each individual house, with such water as the public service did not need. In Silchester the water arrived below the town level and was raised with force pumps. An 8-mile aqueduct supplied the gold workings at Dolaucothy (Carmarthenshire) with the necessary water to wash the ore.

Financing the aqueduct was essentially a state affair. The land on which the aqueduct was to be built had to be bought by the state or the town council. As no rights of expropriation existed this was often a difficult affair and the entire property would be bought and the land not needed would be resold. A strip of 12 Roman feet was usually reserved and even then this strip was leased for pasture, cutting grass and hay or cutting brushwood. The earlier aqueducts were often financed from booty, that is from funds robbed in foreign countries. Then the state shifted this burden onto the shoulders of "viri triumphales", they became a benefaction of succesful army-commanders. Finally the state tried to extract funds from the taxpayer. Both in Rome and in many provincial towns a watertax was imposed on private and business houses that piped water from the main aqueduct. At Rome the public could draw water freely without paying and even by the end of the first century A.D. the water-tax collected amounted to some 250.000 sesterces per year, hardly enough for the upkeep and repairs. Italian cities usually managed to pay for an excellent supply of clean water. Villages paid water-tax to a neighbouring town for the use of a branch line (80). In Roman Syria the temple officials managed to finance aqueducts out of the income of the temple. All over the Roman Empire public-spirited men like Sextillius Pollio of Ephesos managed to offer them as a gift to their home-town honoris causa.

Lay-out and organisation of the Roman water-supply

Classical authors inform us more particularly of the water supply of the city of Rome apart from their general discussion of the problem. With the words "but let us now discuss marvels which, if properly estimated, are unsurpassed" Pliny (81) introduces a discussion of the properties of water, the remedies derived from mineral waters, how to look for the signs indicating its presence and how to convey it. Unfortunately his account is rather superficial and only a few lines are informative. Vitruv devotes the entire eighth of his ten books on architecture to water (82) summarizing the indications, properties and tests of water, aqueducts, wells and cisterns and the levelling instruments used in their construction. His observations, accurate as they are, should not be taken to represent actual data on Rome's water supply in his days. He hardly mentions the use of siphons and the supply of water to baths, fullers, workshops and other industrial establishments, nor does his description of town installations like castellae agree with archaeological data. His essay is rather a vision of a well centralized water-supply of the future in a period of transition from republican collective power to the dictatorial power of the emperors. It was suggested (83) that many of Vitruv's ideas were adopted by Augustus as shown by the water regulations of Venafrum (84) and by Agrippa's reorganisation of the water-board of Rome.

Our main information comes from Sextus Julius Frontinus (85) thrice consul, imperial legate in Britain (where he won a victory over the Silures of Wales (86)) and water commissioner of Rome from 97 A.D. to 103/104 A.D., whose two books on Rome's water-supply abound with practical technical information. He too is much impressed by these great public works exclaiming: "With such an array of indispensable structures carrying so many waters, compare, if you will, the idle Pyramids or the useless, though famous, works of the Greeks!" (87).

Originally the water-supply of Rome was controlled by municipal magistrates (censores and aediles). Augustus centralised these duties which he entrusted to his friend Marcus Vipsanius Agrippa. His regulations were confirmed by decisions of the Senate 11—9 B.C. and laid down in a law of 9 B.C. This entrusted a board of curatores with these duties, which were controlled by an Imperial procurator aquarum. This board consisted of a chairman of consular rank (curator aquarum, Water Commissioner) and two technical advisors (adiutores). Frontinus was the seventeenth Water Commissioner of Rome, Agrippa the first. During the reign of Claudium a reorganisation (52 A.D.) entrusted the procurator aquarum with the entire responsibility and the necessary funds for upkeep and construction.

The Water Commissioner disposed of two lictores (beadles, who were never active), three servi publici (officials), an architect and an administrative staff (scribae, librarii, accensi and praecones). His technical personnel consisted of villici (who attended to the pipes and orifices), castellarii (for the reservoirs), circuitores (line-inspectors), silicarii (paviours) and tectores (masons). Agrippa had used a band of 240 slaves of his own, whom he trained and left to the state at his death and Claudius had formed "Caesar's gang" of 460 men. These gangs were then combined and together with the specialists mentioned above, further architects (aquarii) and free labour (hired occasionally) they formed the regular technical staff.

This large staff was certainly needed for out of the 95 larger aqueducts of the Roman Empire (88) eight large ones served the capital. Five aqueducts drew their waters from springs and artesian wells, two drew on river water and one sea-water. The older aqueducts were mainly subterranean, an excellent precaution against war risks. The later ones reached the city at the highest possible level and thus risked being cut as happened during the siege of Rome by the Goths. Five aqueducts could deliver to all city levels. The total section of all these ducts was 7.587 m^2, that is the same troughput as a cylindrical main of 3.11 m^2. The 351.6 km length of aqueducts (only 47.4 km overground!) delivered about 1.010.623 m^3 per day (VI) against 286.000 in 1886.

In Table X these figures have been converted to consumption per head per day, but it should be remembered that ancient life demanded many outdoor amenities which we never had. Thus in the fourth century A.D. Rome had 11 public baths (thermae), 856 private smaller baths and 1352 fountains and cisterns. This enormous supply, the many public services, the measuring orifices and branch lines, and the city reservoirs (castellae) certainly demanded the staff discussed above.

When the source of water (spring or river) had been chosen plans for its future course were carefully made. Underground ducts were preferred and the beautiful ducts on tiers or arches were avoided because of their high cost. Plans were drawn after a course had been chosen, Frontinus mentions that he had carefully drawn plans of all existing aqueducts and lines. The course chosen usually had a gradient of 1 : 200—1 : 1000 (though we sometimes find one of 1 : 3000). Every attempts was made to avoid a natural creation of a pressure-supply, for siphons were expensive to build and repair. The siphon was occasionally used to cross sharp valleys and to bring the water back to its own level but in no sense do we find an anticipation of the modern high-pressure supply. In pumps and waterorgans hydraulic pressure was sometimes used by the Roman engineer but not in water-lines.

TABLE X

Public Water Supply in Gallons per head per day

	50 B.C.	A.D. 100	1823	1830	1835	1936
Rome	198	300		250		150
Paris			3			
London			3		10.0	35.5
Manchester					5.5	33
Liverpool					3.5	36.5
Edinburgh					7.5	52
Glasgow					12.0	57
Leipzig						20
Frankfort						40
Münich						55
New York						120

The gradient of the aqueducts was not always evenly maintained. This was due to the simple levelling instruments available. The Ancient Near East had known a plumb-bob level (in the form of capital A with a plumb-line hanging from the apex) and sighting devices such as the merchet. The Greeks and Romans had the libra aquaria (a simple water-level), the chorobates (a primitive plane table or field service level some 20′ long), the dioptra (a waterhorizon with sights mounted on a divided circle), the groma (or Grecian star, to trace 90° angles) and levelling rods (90). Skilful work with these primitive instruments would, however, enabled them to maintain a gradient of 1 : 2000 very closely. Even accurate tunnelling was possible with these instruments and only seldom do we hear complaints of careless and negligent work such as those of Nonius Datus on the deviations of the water-tunnel of Lambaesis (Mauretania) in 152 A.D.

From spring or river the water was led to settling tanks (piscinae, castellae limonariae), which usually had two compartiments with sloping floors, which facilitated cleaning. The water was then conducted to the real duct (specus), often 50′ underground, which had air-shafts (lumina, columnaria) every 40—50 yards (av. 1 actus = 120′) to prevent air-locks, and to allow inspection and cleaning. The area of the specus varied from 0.5—3.0 m², it consisted of a 50—60 cm concrete lining enclosed and carried by a mass of masonry. Modern aqueducts avoid this to prevent cracking or stresses in the masonry because of temperature differences between the water and the aqueducts. As many of the springs tapped were fairly hot the Roman aqueducts suffered from serious deposits and incrustations of carbonate of lime and had to be cleaned out fairly frequently. This cleaning was done by the circuitores but even now these incrustations mark the course of the aqueducts.

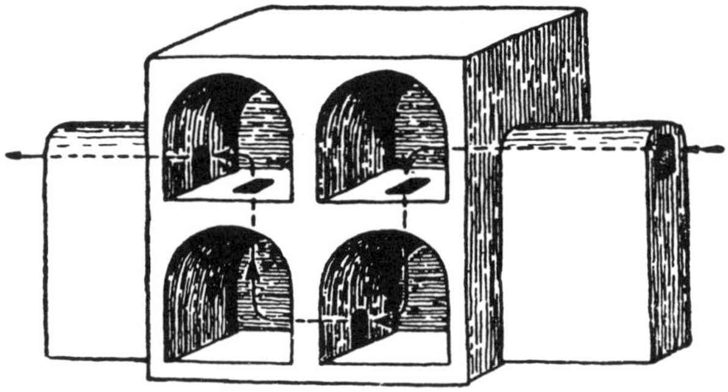

Fig. 37.
Cross-section of a settling tank of the Aqua Julia (Rome).

Only the waters of the Aqua Virgo and Aqua Marcia were so pure that they did not even need piscinae (91); that of the Aqua Alsietina was undrinkable and with part of the supply of the Aqua Traiana served for water-mills, naumachia (water-shows) and sewer-flushing.

The water of the aqueducts arrived at the castellae, which were not reservoirs in the modern sense of the word as they had only little storage capacity. As the Roman watersupply worked on the principle of constant offtake they were distribution tanks only. Each aqueduct had a number of such castellae; the Aqua Appia served 20, the Anio Vetus 92 castellae, the total for Rome amounting to no less than 247.

Town installations and distribution

According to Frontinus there were three large groups of consumers (92): 1) those to whom the emperor had granted supplies (sub nomine Caesaris) which included several public services, 2) private parties, and 3) public supplies.

If we summarize the figures given by Frontinus we get

At the emperor's disposal (incl. baths)		17.1%
Private parties (houses and industries)		38.6%
Public supplies:		
19 military barracks	2.9	
95 official buildings	24.1	
39 public buildings & theatres	3.9	
591 cisterns & fountains	13.4	44.3
		100.0%

From the castellum three mains tapped water near its bottom for supplies to the fountains, thermae (baths) and official buildings. Ten higher mains (which therefore had less head) delivered water to private consumers, blocks of houses, industries, etc. Though tapping from the specus or main was unlawful, the overflow from reservoirs, fountains and public baths (called "lapsed water" or "aqua caduca") was often free though it was also used for swilling drains or for industrial purposes. The basement of the baths of Caracalla contained mills driven by this surplus. Apartement houses and other buildings had secundary reservoirs in which the water supplied was pumped by water-wheels, force-pumps and the like.

Vitruv devised a way of balancing the demands of the groups of consumers during every period of the day which he describes thus (93): "To the castellum a triple receptacle is to be joined, to receive the water; and three pipes of equal size are to be put in the castellum, leading to the adjoining receptacles, so that when there is an overflow from the two outer receptacles, it may deliver into the middle receptacle. From the middle receptacle pipes will be taken to all pools and fountains; from the second receptacle to the baths, in order to furnish a public revenue; to avoid a deficiency in the public supply, private houses are to be supplied from the third... The reason why I have made this division, is in order that those who take private supplies into their houses may contribute by the water rate to the maintenance of the aqueducts."

This proposal was not generally applied as each quarter of the city had its own range of uses for the water supplied. The type of castellum described by Frontinus was usually well-fitted to cope with the demands at any time of the day. In provincial towns supplies were not always so plentiful and the authorities like those of Pompeii were often to ration or cut off the supplies to private persons during certain periods of the day in order to dispose of sufficient water for the baths and public buildings which formed the hub of town-life.

The emperor could grant any syndicate or person (even for life) the right to tap the mains for his own use but generally the aquarii who were in charge of each castellum delivered water to consumers and charged them according to the only standard known then, a nozzle or ajutage, taking the maximum throughput per day of such a nozzle as a basis for their calculations. We should realize that the Roman engineers had only the faintest ideas of hydrodynamics. They were of

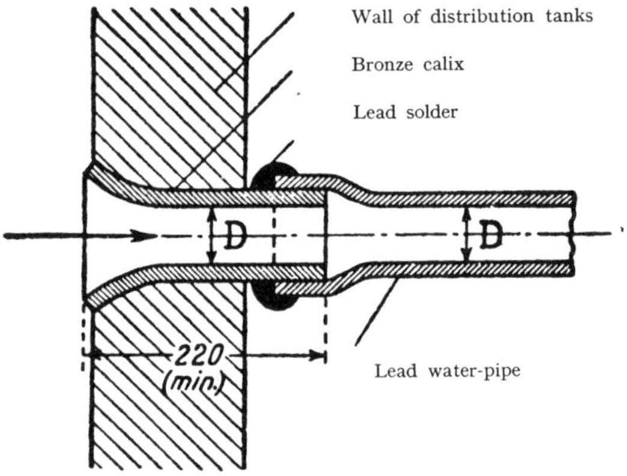

D: diameter of delivery-pipe (fistula)

Fig. 38.
Calix according to Frontinus.

course conversant with the rule expressed by Pliny in these words: "Water, it should be remembered, always rises to the level of its source." But he also adds: "If, again, it is conveyed from a considerable distance it should be made to rise and fall every now and then, so as not to lose its motive power" (94). The influence of the height of pressure, the grade, the resistance of the channel or pipe and the velocity of the water was not properly known as will be seen from this passage from Frontinus (95): "Every stream of water, whenever it comes from a higher point and flows into a reservoir after a short run, not only comes upto its measure, but actually yields a surplus; but whenever it comes from a lower point, that is, under less pressure, and is conducted a longer distance, it shrinks in volume, owing to the resistance of its conduit; and that, therefore, on this principle it needs either a check or a help in its discharge".

The ajutage or calix took the place of our watermeter and based on a certain maximum possible throughput given a certain average head of water one was charged independent of the individual offtake or the head at the specific spot of the main where the calix was tapping its water. For "the calix in a bronze ajutage, inserted into a conduit or reservoir and to it the service pipes are attached. Its length ought not to be less than 12 digits, while its orifice ought to have such capacity as is specified (in any particular instance). Bronze seems to have been selected, since, being hard, it is more difficult to bend, and is not easily expanded or contracted." (96).

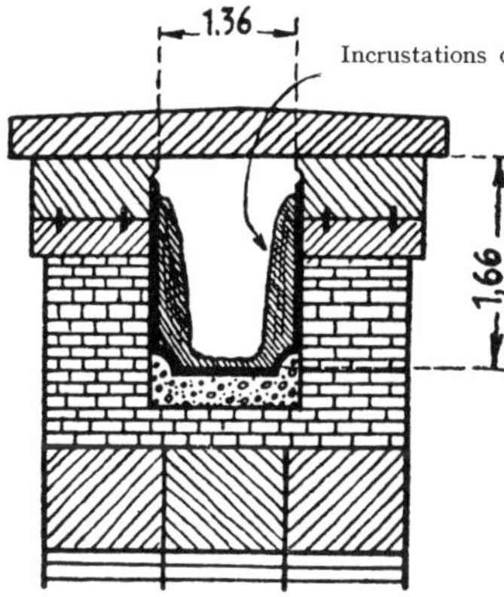

Fig. 39.
Cross-section of the Pont du Gard duct (Nîmes).

Frontinus was probably vaguely aware of the properties of a Venturi-tube, where there is a considerable drop of pressure in the orifice without loss of pressure beyond it (97) as he says: "Placed at right angles and level, the calix maintains the normal quantity. Set against the current of the water and sloping downward it will take in more (devexus ampluis rapit). If it slopes to one side, so that the water flows by, and if it is inclined with the current, that is, is less favourably placed for taking in water, it will receive the water slowly and in scant quantity."

These calices delivering to the consumers through the private pipes (fistulae) were placed in such a way that no more than 1 quinaria of water was taken off every 50′. The unit of the calices was the quinaria (both a standard of the area of the pipe (m^2) and its throughput (m^3/h)), the "five-digit" lead pipe, discussed above, which had an area of 4.453 cm^2. Both Vitruv and Frontinus discuss the set of the standard pipes based on the quinaria (98). The latter author details a series of 25 adjutages ranging from a "one-digit" to a "120-digits" pipe, 15 of which were in common use only. As these standard ajutages were the only ones allowed to the consumer, we have here the first logical series of standard parts in the history of technology, which were carefully inspected and stamped by the villici. "Care should be taken, as often as an ajutage is stamped, to stamp also the adjoining pipe over the length prescribed (at least 12 digits). For then only can the overseer (villicus) be held to his full responsibility, when he understands that none but stamped pipes must be set in place."

"In setting ajutages also, care must be taken to set them on the level and not place the one higher and the other lower down. The lower one will take in more; the higher one will suck in less, because the current of water is drawn in by the lower one." (99).

Though earthenware pipes (tubuli) were sometimes used for private pipes lead was more common. The lead pipes were made from strips, the edges of which were brought together and soldered. As the solder was often pure lead, tin being expensive, leaks were rather frequent. The lead was usually cast in 10′ sheets and cut up into strips of the correct width. If private reservoirs or cisterns were tapped lead pipes were also used, but all fittings, stops and faucets, of which the Roman world knew many types, were made of bronze. Some of the private houses in Pompeii have as many as 30 taps. The city had but few mains and the private citizens often tapped their water directly from the castellae and conveyed it over long distances.

Despite the large quantities of lead pipes frequently needed a factory system of manufacture failed to emerge. A tenacious smallshop system resisted the obvious economic inducements towards centralisation. Part of the work of the slaves of the water board consisted in making and laying lead pipes for the public services. Like other pipes they bear the name of the maker, the Water Commissioner and the emperor as well as the date. But the board also employed the services of private plumbers. Larger numbers of pipes were contracted for by private contractors (or well-to do) who had purchased the right to tap the public water-mains. Their pipes are regularly stamped with the owner's name for identification as well as that of the contractor. From a survey of these it seems that there were few really big contracts, one seldom finds one name in two widely separated regions of the city. It would seem that small owners of plumbers' shops with a few slaves took orders as they came, bought materials, rolled them into sheets from which the requisite strips were cut, pipes made and soldered and finally laid and connected.

Frontinus not only surveyed and restored the whole water-supply of Rome, he also tried to establish correct data on which his bureau could work. He was not satisfied with the total intake of the aqueducts of 12.755 quinariae mentioned in the imperial records but started to measure the quantities of water delivered at the sources, at the reservoirs and finally to the consumers. This showed him that at the intake 18.433 quinariae went to Rome and that finally 14.018 were delivered. Due to his lack of knowledge of sound hydraulics he imputes the loss entirely to leakage or to thefts by tapping without ajutage, widening existing ajutages or using unchecked, too wide ones. This he corrected and on this account alone he was justified in saying: "The expense of a monument is superfluous; my memory will endure if my actions deserve it." (100).

Testing and purifying water

Both ancient doctors and engineers have stressed the necessity of obtaining pure water. Hippocrates, Galen, Vitruvius and many more have denounced the use of lead for cistern-linings or pipes and this frequency of lead poisoning has recently been repeated (101). However, the manifold cases of rather intense incrustations of calcium carbonate on the interior of ancient lead pipes suggest that this complaint is grossly overrated.

Water should be properly tested and Vitruv apart from observing the "physique of those who live in the neighbourhood" recommends the following methods (102): "The water, being sprinkled over a vessel of Corinthian bronze (a gold-silver-copper alloy) or any other good bronze, should leave no trace. Or if water is boiled in a copper vessel and is allowed to stand and then poured off, it will also pass the test, if no sand or mud is found in the bottom of the copper vessel. Again, if vegetables, being put in the vessel with water and boiled, are soon cooked, they will show that the water is good and wholesome." The spring of the water should not be "defiled with filth" neither should "moss nor reeds" grow there. The practical boiling test with vegetables was already mentioned by Hippocrates who states that waters "with a very solvent nature" are preferable for cooking in order to stimulate digestion. Boiled vegetables should not be "hardened" by the water.

Several ancient texts mention that certain waters "can bear a little wine" and it is quite possible (103) that by dosing water drop by drop with a strongly coloured wine (vitis faecenia) of the Algerian type the Romans were able to estimate roughly the lime content of their water.

There were many ways of purifying water (104) such as the very ancient ones of exposing it to sun and air or boiling it like the Persian kings who drank water of the Choaspes kept in silver flaggons (105), a method sponsored by Hippocrates. Others like Aristotle (106) prefer filtration through porous pottery made of mixes of zeolite and clay, or of wax, as some prefer to read (107), however dark the text may then be. However porous filters made of tufa have been found in several places (108). Filtration through wool or wick syphons was well-known. Athenaeos of Attalia wrote a *Book on the Purfication of Water* (50 A.D.) in which filtration was discussed and natural desalting by percolation of sea-water into certain subterraneous galleries in Egypt commented upon. Percolation through layers of sand is also recommended by Vitruvius (109). "If cisterns are double or treble so that they can be switched, they will make the supply of water much more wholesome by percolation. For when the sediment has a place to settle in, the water will be more limpid and will keep a flavour unaccompanied by smell. If not, salt must be added to purify it." The addition of salt is also recommended by Palladius if the water contains too much lime. It was in fact an old remedy for "Elisha went forth unto the spring of the waters (at Jericho) and cast the salt in there... So the waters were healed unto this day." (110).

Certain herbs were also known to purify water since "the Lord shewed Moses a tree which when he had cast it into the (bitter) waters (of Marah) the waters were made sweet" (111). The *Geoponica* mentions barley in a bag, macerated laurel or bruised coral and Pliny advises to cook the water for two hours with polenta or to use chalk of Rhodes or clay (argilla) from Italy. It was also believed that "bronze has a purifying effect" (112) and hence the automata for holy water were made of this metal. The most common remedy, however, was mixing water with wine as is still the general custom in Southern Europe.

Water-supply in the Middle Ages

With the fall of the Roman Empire the lack of central authority and large public funds led to the decline of all such public services as road-building and watersupply. The planning and reorganisation of such services was left to either groups of private citizens or municipal authorities. Only in urban centres of Roman date did they survive to some extent. The new cities of Western Europe were thrown back on more primitive means.

As early as 389 A.D. a law had to forbid the landowners to tap the aqueducts to irrigate their lands, and around 600 the bishops were want to take charge of the aqueducts to ensure a public supply of water. In most cases the town people were thrown back on wells, springs and rivers for their supplies. As care for proper sewers and street-cleaning had seriously declined (113) hygienic conditions were more closely connected with water-supply than during the Roman Imperium when pure water was obtained far beyond the bounds of the city. The archbishops of Salzburg obtained their daily supply of fresh spring water by messenger, but the ordinary citizen had to rely on his private well or cistern or on the town wells and fountains.

Wells, being also used for the storage of rain water, were often too close to cesspools and latrines. Epidemics spread quickly. In the town of Strassburg thousands died in the fourteenth century because of such contaminated wells (114) whereas the Jews in that town suffered little as their doctors had forbidden them to drink this water and told them to rely on boiled river-water or rain-water. None of the many municipal laws on defiling the rivers could stop this process. Open gutters in the middle of the street into which refuse was thrown and rain water poured from the roofs were not properly drained in sewers. The streets, seldom paved, were often mud-pools in which pigs crowded and their water leaked into the wells or private plots.

These conditions slowly changed when the local authorities grow less indifferent to street-cleaning and sewers. In the course of the fourteenth century the Flemish towns set the good example (115) and also enforced laws on clean water and food. The Black Death stimulated such measures like many others, it was the cause of the first Urban Sanitary Act of 1388 which forbade throwing filth and garbage in ditches, rivers and waters in all England.

The water from public fountains and wells was often distributed by professional water-carriers. In Paris, Alexander Neckam found in the twelfth century (116) each house-owner kept his water in a large open vat ("tine") which was kept filled by the water-carrier who had contracted to do this job. The water was carried by him in two wooden pails with the help of a yoke ("grouge"). As most of the Paris wells gave brackish water, much water was taken from the river Seine, which then flowed quicker and hence was cleaner. The town wells worked either with a bucket and windlass or if water was near the surface buckets with a counterweight were used. Many towns changed over from wooden buckets to copper ones because they could be more properly cleaned.

Things started to change when Philipe Auguste enclosed Paris with

Fig. 40.
Roman aqueduct of Segovia (Spain).

walls about 1190. At that time the abbey of Saint-Laurent had a water reservoir at Pré Saint-Gervais which held water from springs at Romainville. This water was piped to Paris in lead pipes (117). The abbey of Saint-Martin des Champs had repaired 1200 yards of masonry of the old aqueduct of Belleville and mains were now being laid to the public fountains of Paris. The Convent of the Filles-Dieu was the first to get a concession to pipe water from the public main to its establishment (1265) and many nobles and merchants followed suit. However, illegal tapping became so frequent that a special decree (Oct. 9, 1392) had to be enacted. In 1404 Charles took the necessary steps to prevent pollution of the Seine. Citizens began to take an interest in proper water-supply and the Provost of the Merchants paid the repairs of 200 yards of the Belleville aqueduct (1457). By that time things were on the move and pumping from the Seine with the help of a water wheel was discussed, though these plans were not realised for some time to come.

What was related here about the developments in Paris is typical for the story in many medieval towns. Some of them were able to keep the old Roman aqueducts more or less in repair. Others like towns in the west of England (Plymouth, Devonport) could build a primitive kind of aqueduct (or "leat"), carrying water from a spring or river into the town in an open duct. A few towns like Konstanz were able to pipe mountain springs to town. The advent of pumps in the fifteenth and sixteenth century changed the picture completely. Attemps at pumping started earlier but failed to become common because of the primitive machinery. Undershot water-wheels, some 5—6.5 m in diameter, driving piston-pumps were being introduced by the end of the fifteenth century in south German towns. The "Seven Children" fountain at Augsburg (1450) pumped up water with a series of Archimedean screws.

Wooden pipes were preferred in many parts of medieval Europe. Early in the nineteenth century one third of Augsburg's water-mains (3000 m) were still wooden. These wooden pipes were usually 6 m long and 8—12 cm in diameter, working at pressures upto 3.5 atm., but generally much less. Some towns like Nürnberg which had an elaborate system of pipes even had a special drilling machine for wooden pipes illustrated in a mss. of 1430. In 1462 Endres Tucher replaced it by a more efficient one which he shows in his *Baumeister Buch* (1470).

Pottery and lead pipes came in the second place only. The lead pipes (usually 2″) were soldered with tin solder and held very well judging from the archaeological finds. Still Vincent of Beauvais in his *Mirror of Nature* (1254) said that tin solder facilitated corrosion and that lead should be used like the Romans did. Cast iron pipes were used for the "Wasserkunst" (public fountain) of Augsburg in 1412 but after four years they were exchanged for wooden ones. Calices were still used in some towns and even in English leats "oxe-eyes" (stones or slates with an orifice) were used (118).

In the early Middle Ages ecclesiastical authorities and monasteries were pioneers in water-supply. Later municipal authorities or groups of citizens took over this task but law suits between the two were frequent. The cost of driving wells was born by the town or groups of "well-brethern". The latter usually catered for their members and left the problem of supplying water for the poor to the town, clamouring when money was exacted from them for some public service. The actual sinking or repair was left to professional well-diggers ("Kunstmeister", "Brunnenmeister") often in reality lay-men, who first set out to examine similar works in neighbouring towns. Only when a fairly large public water-supply existed, was a special surveyor or caretaker appointed.

By the end of the Middle Ages scientific works on water-supply begin to appear. Konrad Keyeser devoted the whole third book of his *Bellifortis* (1605) to hydraulics and water-supply. His contemporary Giovanni da Fontana not only wrote on aqueducts (1420) but actually proposed a hot air engine to raise water from a well. They are the heralds of a new dawn, the era of the pump which transformed water-supply from a gravity-flow system to a pressure-system.

Chronological survey of the story of water-supply[1])

c. 2500 B.C.	Indus valley cities build steined wells and drainage systems, Mesopotamian cities follow suit.
c. 2500	Lead stoppers and copper drain-pipes found in the mortuary temple of Sahurê (Abusir). Hand-dug wells built in Egypt and Mesopotamia. Drainage systems common in urban life. Collection of rain-water in cisterns spreads.
c. 2000	Conduit for spring water built for Palace of Cnossos (Crete). Rain water stored and drainage system installed.
c. 1900	Subterranean tunnel (ṣinnōr) to spring outside the fortifications constructed at Gezer (Palestine).
c. 1500	City of Tell Ta'annek (Palestine) connected with two springs by tunnel (ṣinnōr).
c. 1300	Sinnōr of Megiddo built in stages (5 × 5—7 m; 82 m long).
c. 1200	Mycenaeans the first Europeans to sink wells and to build elaborate drainage systems. Water from Perseia spring brought to Mycenae by conduit (400 m) and tunnel (100 m).
1055	Drains in Jerusalem.

[1]) For most aqueducts the total length, length of the overground portion and area of the specus are given.

c. 1000	Solomon's cisterns built in Jerusalem. Springs encased in planks common in Bronze Age Europe, the spring at St. Moritz encased.
870	Assurbanipal has wells dug at Niniveh.
c. 850	The Hallstatt civilisation of Central Europe starts collecting and storing rainwater.
c. 800	The Etruscans built narrow-mouthed bell-shaped steined wells. Watermains of Van (Armenia).
714	Sargon II destroys the qanāt-fed irrigation systems of King Ursâ of Urartu at Ulḫu (Armenia).
703	18 settlements near Niniveh supplied with water from upper Khosr river by a 10 mile canal and a weir near Kisri.
c. 700	Hezekiah's water-tunnel to the spring of Siloah (8′ × 8′), a conduit (537 m long, fall 0.35) brings water to bottom of city-shaft. Metal water-lines installed at Tyre. Water-cisterns of Thera (Greece). Conduit brings water of the upper Thymbrius to Troy (7 miles).
694/690	Sennacherib's aqueduct from Mount Tas (Bavian) to Niniveh (55 km).
c. 600	Theagenes builds conduit for Hymettus water to Athens for Pisistratos. Conduit of spring water built at Megara.
594	Solon provides for a well on every farm.
590	Tarquinius Priscus the Etruscan builds Rome's Cloaca Maxima (14′ × 11′).
c. 580	Nebuchadnezzar II builds qanāt-like drains of the base of the ziggurats of Borsippa; same type of drain under ziggurat of Ur.
c. 530	Eupalinos builds water-supply of Samos (including a 1100 m tunnel) for Polycrates.
c. 500	The Persians open up water supply for Kharga Oasis by qanāts.
491/477	The Crimiti water-conduit for Syracuse (Sicily) built.
489/472	Aqueduct of Acragas (Sicily).
c. 450	Aqueduct of Girgenti (Sicily).
332	Aqueduct and cloaca of Alexandria (Egypt).
320	A 600′ well dug in the Sahara.
312	Aqua Appia Claudia, Rome's first aqueduct (length

	16.56 km, overground 90 m, section 0.74 m^2) (output 75.737 m^3/day).
307	Aqueduct of Antiochia (Syria).
300/200	Aqueducts built in various Roman towns in Italy.
272/269	The Anio Vetus (Rome) built from the spoils taken from Pyrrhus (length 63.6 km, overground 0.327 km, section 0.95 m^2, used for irrigation and drainage only) (output 182.517 m^3/day).
260	Aqueduct of Shuster (Persia).
c. 250	Wooden conduits and pipelines, true planked wells common in northern and central Europe.
230	Greek authors mention bronze taps, manifolds, more-way taps and orifices.
c. 200	Aqueduct of Smyrna gets pressure syphon (leading from + 200 m hill-top through + 26 m valley to cistern in town at + 184 m).
200	Thermae introduced in Rome.
179/159	Eumenes II lays pressure water-line at Pergamon (three 19 cm^2 lines, conduct water from spring at 1200 m to double chamber at Hagios Georgios (+ 360 m), syphon through valley with pressure upto 20 atm. brings water to 5 m square, 8 m deep cistern in city-castle at + 335 m.
c. 175	Water-supply of Pisaurum (Italy) built.
168	P. Aemilius strikes artesian well in Macedonia.
144/140	Aqua Marcia built for 180.000.000 sesterces (£ 3.500.000) (Rome) (length 91.7 km, overground 10.25 km, section 1.18 m^2), later enlarged during Roman Empire (output 194.365 m^3/day).
134	L. Betilienus Varus builds aqueduct with syphon (10 atm.) for town of Aletrium (Alatri, Italy).
127/125	Aqua Tepula provides the Capitol with water (section 0.18 m^2) later joined with Aqua Julia.
89	Sergius Orata invents central heating.
63/13	Aqueduct of Gardon (Nimes, France) (length 49.75 km).
35/33	Aqua Julia (Rome) (length 22.8 km, overground 9.5, section 0.485 m^2, output 50.043 m^3/day).
34	Marcus Vipsanius Agrippa first permanent Curator Aquarum for the town of Rome.

33	Agrippa forms a "familia" of slaves for the upkeep and repairs of aqueducts at Rome.
30/14	Aqua Alsietina (length 32.8 km, overground 0.5 km, section 4.80 m^2) not used for drinking water. (output 16.228 m^3/day).
	Aqua Augusta (length 1184 m, entirely underground).
24	Vitruv writes on water-supply and aqueduct construction.
21/19	Aqua Virgo (length 20.88 km, overground 1.03 km, section 1 m^2, output 103.916 m^3/day).
12	State department of aqueduct repairs, 250—450 men (Rome).
11	Permanent board of curatores aquarum (Rome).
11 A.D.	Augustus builds aqueduct from Schedia to Alexandria (Flumen Augusti; 40 km).
35/49	Aqua Anio Nova (Rome) (length 86.8 km, overground 15.0 km, section 1.90 m^2, output 196.627 m^3/day).
38/52	Aqua Claudia (Rome) (length 68.7 km, overground 13 km, section 1.33 m^2, cost 350.000.000 sesterces, output 191.190 m^3/day).
50	The emperor Claudius constructs the aqueducts of Lyon (500—4800 m), with the "pont-siphon of Garon". These contain 18 syphons (levels differing as much as 25—123 m).
	Claudius finances the aqueduct of Sardis, builds aqueduct of Ravenna.
74	Public conveniences of Rome get running water.
77	Pliny describes properties, caption and purification of water.
79/80	Aqueduct of Smyrna.
85	Silver water-pipes in Roman villas.
88	L. Paquedius Festus builds 3 mile tunnel to shorten Aqua Claudia.
97	Sextus Julius Frontinus writes memo on water-supply.
100	Frontinus' book published. Earthenware pipes with sockets used in Rome.
109	Aqua Traiana (length 57.8 km entirely underground, section 3.02 m^2), later largely used for power (water-

	mills). Aqueduct of Segovia (Spain) (length 15 km), contains arched bridge 2600′ long, 90′ high.
117/138	Aqueduct of Athens built by prominent citizen.
123	New aqueduct of Carthage (Zoghouan), output 6.000.000 gls/day.
c. 125	The emperor Hadrian construct the aqueducts of Arles (France, 67 km), Corinth, Antioch, Sarmizegethusa (Austria), Dyrrachium (Yougoslavia), Gabii, Naples and Cingulum.
130	Aqueduct of Metz, output 40.000 m^3/day.
148/52	A 428 m water-tunnel excavated for Lambaesis (Bougie, Algeria).
160	Aqueduct of Olympia (Greece).
222/235	Aqua Severiana (Alexandrina) (length 22 km, overground 8 km, section 1.25 m^3).
250	Siphon in branch of Aqua Claudia replaced by conduit on 4 tiers of arches, Aqua Marcia repaired. Aqueduct from Eiffel to Cologne (85 km).
305/306	Conduit of earthenware pipes brings water from Arceuil to the Roman baths near the Palais Royal (Paris), 117.000 gls/day.
360	The emperor Julian builds the "aqueduct de Rungis" for Paris, output 2600 m^3/day.
366	The emperor Valens finishes the aqueducts of Constantinople.
528	Justinian builds the underground cistern "Hall of the 1001 Columns" (Constantinople) (70 × 60 m surface) to store water supplies from the Thracian hillside.
537	Aqueducts of Rome cut during siege by the Goths. Floating water-wheels invented.
541	Theodoric builds aqueduct of Spoleto with a 277 km bridge 145 m high.
550	Justinian rebuilds Hadrian's aqueduct of Constantinople.
776 A.D.	Pope Hadrian I restores several Roman aqueducts (Traiana, Marcia, Claudia, Virgo).
1030	Deep well dug at Nüremberg (Germany).
1126	First Artesian well in Western Europe dug at Lillers (Artois, France).

1150	Cistercensians use refuse water to fertilize meadows (Italy).
1160	Canterbury cathedral and close supplied by conduit made from plan by Prior Wibert.
1183	Water-supply of Paris improved.
1200	Conduit built for town of Frankfort, drainage system installed.
1226	Water from St. Sidwell's spring conducted to Exeter cathedral close.
1236	Gilbert Sandford builds first London public conduit from Tyburn.
c. 1250	Water-supply of Dublin built "at the cost of the citizens".
1273/1279	Aqueduct of Orvieto.
1300	Thurn und Taxis builds drains of Regensburg, Hamburg drainage system constructed.
1320	The Paris "aqueduct de Belleville" already in use.
1349	Deal and fir water-pipes used in Germany.
1370	The Carlsberger mineral spring discovered.
1388	Public Water-supply of Nüremberg; cloacae of Vienna built.
1400	Aqueduct of Wismar (Germany).
1430	Spring-water used for public water-supply Zürich. Charterhouse conduit (London) built.
c. 1450	Introduction of cast iron water pipes in Germany and Great Britain.
1457	Part of Belville aqueduct rebuilt.
1471	Aldermanbury conduit (London).
1478	Fleetstreet, Fleetbridge, Cripplegate conduits (London).
1491	Grass Market cistern and Snow Hill conduit built (London).
1500	Aqueduct Kolberg (Germany).
	Suleiman the Great reconstructs the aqueducts of Valens and Justinian and adds one of his own (Constantinople).

Notes

I. Arabic "qanāt" comes from the root " qnī".
The Persian equivalent "kārīz, kahrīz" is derived from "rixtān" (conduit).

II. Arab. "moghani", comes from the same root as "qanāt".

III. Sargon speaks of blocking (sakāru, ÚŠ) the outlet (arūru) of a canal (ḫirîtu, TÚL), the latter terms may denote the qanāt itself.
Nebuchadnezzar calls the "outlet" of his drain "mūṣū (from the wṣ') of water (mē), but the Akkadian synonym list "D" informs us that this expression "mūṣū mê" is synonym to "arūru".
The ḫirîtu is not only the underground part of the qanat but also the canal (duct) from which the smaller irrigation channels sprang.

IV. The name of this town is probably connected with the verb "sakāru" (ÚŠ) (to dam or endyke a river) and the noun "kisirtu", dam, weir.

V. The engineers mentioned in the text were "itunni" ([lú]SITIM) (mason, architect) as contrasted with the "sikiru" ([lú]SITIM-ÍD-DA) (canal-repairers); see S. Langdon, *Stud. Orient.*, vol. I, 1927, 100—101 and B. Meissner, *Lexicographisches* IV, M.V.Ae.G. vol. 15, 1910, 517.

VI. This figure given by Ashby is based on Di Fenizio's calculations who takes a quinaria to give 41.5 m^2. Our knowledge of the pressures and velocities in the Roman water-system is very deficient. The estimate given by Ashby takes 1.5 m/sec. to be the average value, but Kretschmer has recently suggested that 2 m/sec. is a better value. This latter figure would increase the total water-supply mentioned (98).

Bibliography

1. Clark, G., *Water in Antiquity* (*Antiquity*, vol. 18, 1944, 1—15)
Dupont, G., *L'eau dans l'Antiquité* (Paris, 1938)
Buffet, B. et Evrard, R., *L'eau potable à travers les âges* (Liège, 1950)

2. Heierli, J., *Die bronzezeitliche Quellenfassung von St. Moritz* (*Anz. Schweiz. Altertumskunde* N.F. IX, 1907, 265—278)
Lienau, M. M., *Die Quellenfassung von St. Moritz* (*Mannus Z.*, vol. X, 1908, 25—30)
Squassi, Fr., *L'arte idro-sanitaria degli antichi. Epoche preromana e romana* (Toletino, 1954)
Guitard, E. H., *Le prestigieux passé des eaux minérales* (*Soc. Hist. Pharm.*, Paris, 1953)
Gose, E., *Der Trierer Römersprudel, eine Heilquelle aus Römerzeit* (*Trierer Z.*, vol. XX, 1951, 85—95)

3. Simons, J., *Jerusalem in the Old Testament* (Leiden, 1952, 157)

4. Bräunlich, E., *The well in Ancient Arabia* (Leipzig, 1926)

5. *History* III. 6
CHRISTOPHE, LOUIS-A., *Les porteurs d'eau de Deir el Medineh pendant le règne de Ramses III* (*BIE* 36, 381—408)
6. BROMEHEAD, C. E. N., *The early History of Water-Supply* (*Geogr. J.*, 1942, 142—151, 183—196)
7. ROBINS, F. W., *The Story of Water Supply* (Oxford, 1946)
8. NINCK, M., *Die Bedeutung des Wassers im Kult und Leben der Alten* (Leipzig, 1921)
BURROWS, E. E., *The use and worship of water among the Romans* (*Art and Archaeology*, vol. 30, 1931, 221—228)
DUPONT, G., *L'eau dans l'Antiquité* (Paris, 1938)
DRIVER, G. R., *The "Waters of Bitterness"* (Syria vol. XXXIII, 1956, 70—78)
REYMOND, PH., *L'eau, sa vie et sa signification dans l'Ancien Testament* (*Vetus Testamentum Suppl.* VI, Leiden, 1958)
9. VITRUVIUS, *De Architectura* VIII, c. 1 & 2 (edit. GRANGER, *Loeb Classical Library*, 1934)
10. JACOBI, H., *Die Be- und Ent-wässerung unserer Limeskastelle* (*Saalburg Jbb.*, vol. VIII, 1934, 32—60)
11. ii Chron. 26. 10
12. LUCKENBILL, D. D., *The Annals of Sennacherib* (Chicago, 1924, 110, Col. VII, lls. 45—49)
13. HERODOTUS, I. 193
14. POMP, R., *Wasserhebewerke in Pompeii* (*Technikgeschichte*, vol. 28, 1939, 159—160)
15. HUBBARD, ..., *Neolithic Dewponds and Cattleways* (London, 1916)
16. JACOB-FRIESEN, K. H., *Die Ausgrabung einer urgeschichtlichen Zisterne bei Algermissen* (Nachrichtenbl. Niedersachsens Vorgesch. 1925, 35)
17. Jer. 2. 13
18. Isa. 36. 16
19. SCHLIEMANN, H., *Mykene und Tiryns* (Berlin, 1878, 141)
20. FORCHHEIMER-STRYGOWSKI, *Die Wasserbehälter von Konstantinopel* (Wien, 1893)
KRENKOW, E., *The construction of subterran an watersupplies during the Abbaside Caliphate* (Glasgow *Univ. Orient. Soc. Trans*, 13, 1951, 23—32)
21. EGGERS, G., *Wasserversorgungstechnik im Altertum* (*Technikgeschichte*, vol. 25, 1936, 1—13)
22. EVANS, A., *The Palace of Minos* (vol. III, fig. 173) (vol. II, 462—463)
23. BUFFET, B. & R. EVRARD, *L'eau potable à travers les ages* (Liège, 1951)
24. VITRUVIUS, *De Architectura* VIII. 7.
25. MAHUL, J., *Les tuyaux de plomb, histoire et progrès de leur fabrication* (*La Nature*, 1 dec. 1937)
26. PLINY, *Nat. History* 33, 86; 34, 161
FORBES, R. J., *Metallurgy in Anitquity* (Leiden, 1950, 135)
27. PLINY, *Nat. History* 31. 57
28. VON REZORI, H., *Das Gussrohr* (*Gas Wasserfach*, vol. 93, 1952, 295—297)

29. i Chron. XI. 6 (Compare Tudor, D. & Bujor, E., *Die geheime Brunnenanlage von Sucidava* (Dacia vol. IV, 1960, 541—552)
30. Macalister, R. A. S., *A century of Excavations in Palestine* (London, 1925)
31. Lamon, R. D., *The Megiddo Water System* (Chicago, 1935)
32. Starkey, J. L., *Quart. Stat. Pal. Explor. Fund* 1934, 164—175
33. ii Kings 18. 17; Isa. 7. 3; 36. 2
34. ii Kings 20. 20; ii Chron. 32. 30
35. Fischer, J., *Die Quellen und Teiche des biblischen Jerusalem* (*Das heilige Land*, vol. 77, 1933, 41, 89; vol. 78, 1934, 15, 70)
Weill, R., *La cité de David* (Paris, 1947)
Bressan, G., *Il problema del "sinnor"* (*Biblica*, vol. 25, 1944, 346—381)
36. Vincent, L. H., *Jérusalem Antique* (vol. I, Paris, 1912, 158)
37. Reindl, C., *Einiges über den Wasserbau im Altertum* (*Wasserkraft und Wasserwirtschaft*, vol. 34, 1939, 61—65)
38. Thureau Dangin, F., *Une relation de la huitième campagne de Sargon, lines* 199—232 (Paris, 1912)
39. Lassøe, J., *The irrigation system at Ulḫu* (*eight century B.C.*) (*J. Cuneiform Studies*, vol. V, 1951, 1, 21—32)
40. Lehmann-Haupt, C. F., *Armenian einst und jetzt* (Berlin 1910, vol. II, 1. 111)
41. Polybios, *History*, X, 28
42. Strabo, XV. 1.50.c. 707
43. Feilberg, C. G., *Qanaterne, Irans Underjordiske Vandingskanalen* (*Ost og Vest*, Kopenhagen, 1945, 105—113)
de Morgan, ..., *Miss. scient. en Perse*, I. 300
44. Herodotus, III. 9
45. Caton Thompson & Gardner, *The Prehistoric Geography of Kharga Oasis* (*Geogr. J.*, vol. 80, 1932, 373) (see also *J. R. Geogr. Soc.*, Nov. 1932, 382)
Naguib, A., *Subterranean water-conduit at Kom el-Nakla* (*Ann. Serv. Ant. Egypte*, vol. 7, 1906, 95)
Badawy, A., *Au grand temple d'Hermopolis-Ouest* (*Revue Archéol.* vol. XLVIII, 1956, 140—154)
Michalowski, K., *Les filtres d'eau d'Athribis* (J. Jurist. Papyriology XI/XII, 1957/58, 160—166)
46. Vitruv, *De Architectura*, VIII. 6. 3
47. Tablet i.R. 51, 131 see S. Langdon, *Die neubabylonischen Königsinschriften* (Leipzig, 1912, 98)
48. Luckenbill, *The Annals of Sennacherib* (Chicago, 1924, Col. VIII, lls. 20—48)
Jacobsen, T. & Seton Lloyd, *Sennacherib's aqueduct at Jerwan* (Chicago, 1935)
49. Campbell Thompson, R. & R. W. Hutchinson, *A century of Exploration at Niniveh* (London, 1929, 130—132)
Laessøe, J., *People of Ancient Assyria* (London, 1963, p. 103 ff)
Vita-Finzi, Cl., *Roman Dams in Tripolitana* (Antiquity XXXV, 1961, pp. 14—20)

50. JACOBSEN, T. & SETON LLOYD, *Sennacherib's aqueduct at Jerwan* (Chicago, 1935, 6—18)
51. FUAD SAFAR, *Sennacherib's Project for Supplying Erbil with Water* (Sumer, vol. III, 1, 1947, 23—25)
52. SCHAWE, "*Bewässerung*" (*Reallexikon der Assyriologie*, vol. II, 1, 23)
53. LÜTSCHG, O. & LOETSCHER, *Das Wasser* (*Ciba Z.*, vol. 9, 1947, 107, 3893—3940)
SCHMASZMANN, H. & W., *Wasser und Abwasser* (*Ciba Rundschau*, 1950, 91, 3373—3379)
COOK, J. M., *Bath-tubs in Ancient Greece* (*Greece and Rome*, 2nd Ser. vol. VI, 1959, 31—41)
GINOUVÈS, R., *Sur un aspect de l'évolution des bains en Grèce vers le IVe siècle de notre ère* (*Bull. Corr. Hell.* LXXIX, 1955, 135—152)
54. HERODOTUS, III. 60
55. FABRICIUS, E., *Mitt. Deutsch. Archäol. Inst. Athen*, vol. 9, 1884, 165
GLOVER, T. R., *Springs of Hellas* (Cambridge, 1945)
SCHMIDT-RIES, H., *Wasser für Hellas* (Düsseldorf, 1956)
R. S. YOUNG, *An industrial district of ancient Athens* (Hesperia, vol. XX, 1951, 135—288)
56. HERON, *Dioptra, Problems* Nos. 15 & 16 (edit. H. SCHÖNE, Leipzig, 1903)
57. HERON, *Pneumatics, Introduction* (Teubner edit. by W. SCHMIDT, Leipzig, 1899)
DRACHMANN, A. G., *Ktesibios, Philon and Heron* (Copenhagen, 1948, 84)
58. GRÄBER, H., *Die Wasserleitungen von Pergamon* (Abh. Akad. Wiss. Berlin, 1887)
WIEGAND, E., *Die Griechische Hochdruck-Wasserleitung in Pergamon* (*Gas und Wasserfach*, vol. 76, 1933, 513—516)
59. VAN DEMAN, E., *The building of Roman aqueducts* (Washington, 1934)
ASHBY, TH., *The aqueducts of Ancient Rome* (Oxford, 1935)
JANICEAUD, G., *Conduites d'eau* (*Bull. Comm. Yrav. Hist.*, mars 1948, I—III)
60. BERNARDI, M., *L'igiene nella vita publica e nella vita privata di Roma* (Undine, 1941)
61. C.I.L. IX, 5144
62. STRABO, III, 2. 8. cap. 146; PLINY, *Nat. History* 33, 68—77
63. BENOIT, F., *L'usine de meunerie hydraulique de Barbegal* (*Rev. Archéol.* (6), vol. XV, 1940, 19—80)
64. DIODOR, 20. 36, 1—4; LIVY 9. 29. 5
65. FRONTIUS, *De Aquae Ductu* I. 7
66. LIVY, 40. 51. 2—7; 41. 27. 5—12
67. C.I.L. I^2. 1529—30
68. *Res Gestae*, chap. 20
69. PLINY, *Nat. Hist.* 36. 122; SUETONIUS, *Claudius* 20. 1
FRONTINUS, *De Aq.* 14
70. C.I.L. III. 549. 709, 1446; IX. 5681; XIV. 2797; SUETONIUS, *Hadr.* 20. 5; PAUSANIAS, 8. 22. 3
71. DALMAN, J., *Der Valens Aquedukt in Konstantinopel* (Berlin, 1935)
72. GAUKLER, P., *Enquête sur les installations hydrauliques romaines en Tunésie*

NEVEU, R., *L'aménagement en eau potabl des villes de l'Afrique romaine* (*Bull. Soc. franc. hist. méd.*, vol. 18, 1924, nos. 9—10)

CINTAS, J., *L'alimentation en eau de Thysdrus dans l'Antiquité* (Karthago VII, 1956, 179—187)

GODET, R., *Le ravitaillement de Timgad en eau potable* (Libyca vol. II, 1954, pags. 66—72)

73. *I.G.R.R.*, vol. I. 1055
74. ORIBASIUS, I. 337
75. STRABO, 17. 1. 30 cap. 807
76. *Pap. Lond.* 1177
77. WEBER, G., *Wasserleitungen in Kleinasiatischen Städten* (*Arch. Jb*, vol. 19, 1904, 86; vol. 20 1905, 202)

CORMACK, J. M. R., *Epigraphic evidence for the water supply of Aphrodisias* (*Ann. Brit. School Athens*, vol. XLIX, 1954, pags. 9—10)

DIRIMTEKIN, F., *Adduction de l'eau à Byzance dans la région dite Bulgaria* (Cahiers Arch. vol. X, 1959, 217—243)

DOWNEY, G., *The water-supply of Antioch-on-the-Orontes in Antiquity* (Ann. Arch. Syrie, vol. I, 2, 1951, 171—187)

HUBBE, R. O., *Public service in Miletus and Priene in Hellenistic and Roman Imperial Times* (Diss. Princeton Univ., p. 1—50)

PERKINS, J. B. WARD, *The aqueduct of Aspendos* (Pap. Brit. School Rome vol. XXIII, 1955, 115—123)

78. STÜBINGER, O., *Die römischen Wasserleitungen von Nimes und Arles* (Heidelberg, 1910)

DE MONTAUZON, G., *Les aqueducs antiques de Lyon* (Paris, 1909)

MAIURI, A., *Fossa Neronis* (*Bull. Vrienden Antieke Besch.* (*La Haye*), vol. XXIX, 1954, pags. 57—61)

PEZET, M., *L'aqueduc antique d'Arles* (*Bull. Comm. Tr. Hist.* 1946/49, (1953), page 282)

GRENIER, A., *Manuel d'archéologie gallo-romaine* Vol. IV, *Les monuments des eaux, aqueducs, thermes, villes d'eau* (Paris, 1960, 2 vols.)

MARGUIER, J., L' *aqueduc romain de la Bouillide à Antibes* (Provence Historique V, 1955, 26—31)

79. SAMENREUTHER, E., *Römische Wasserleitungen in den Rheinländen* (26° Bericht Röm. Germ. Komm. 1937)

MATSCHOSS, C., *Beitr. Gesch. Technik Industrie*, vol. 17, 1927, 132

SCHMALZ, R., *Der Schlammfang des Kölner Aquaduktes* (*Bonner Jahrb.*, vol. 135, 1930, 105—108)

GOLLOB, H., *Die antiken Bäder von Karnutum und ihr Wiederaufbau* (Wien, 1955)

SCHROT, G., *Wasserversorgung und Kanalisation im antiken Rom* (*Wiss. Z. Unv.* Leipzig, Reihe der Sprachwiss. VI, 1956/57, 285—294)

THOMPSON, F. H., *The Roman aqueduct at Lincoln*(Archaeol. J. vol. CXI, 1954, 106—128)

HABEREY, W., *Neues zur Wasserversorgung des römischen Kölns* (Bonner Jahrb. CLV-CLVI, 1955/56, 156—168)

SCHOPPA, H., *Römische Holzbrunnen in Köln* (Saalburg Jahrb. X, 1951, 76—80)

80. C.I.L. IX. 5144
81. *Nat. History* 36. 101—125
82. VITRUVIUS, *De Architectura*, book VIII (edit. Loeb Classical Library, 1934, vol. 2)
83. GRIMAL, P., *Vitruve et la technique des aqueducs* (*Rev. de Philol.*, vol. 19, 1944, 162—174)
84. C.I.L. X. 4. 842
85. FRONTINUS, *The Aqueducts of Rome* (edit. Loeb Classical Library, 1925)
HERSCHEL, CL., *The two Books on the water-supply of the city of Rome by Sextus Julius Frontinus* (Boston, 1899)
BECK, TH., *Historische Notizen. III. Frontinus* (*Civil-ingenieur* 1887, 343—350)
ASHBY, T., *The aqueducts of ancient Rome* (Oxford, 1935)
DEMAN, ESTER B. VAN, *The building of Roman Aqueducts* (Washington, 1934)
86. TACITUS, *Agricola*, 17
87. *De Aqua Ductu*, I. 16
88. LÉGER, A., *Les travaux Publics aux temps des Romains* (Paris, 1875, 551)
89. *De Aqua Ductu*, I. 17
90. KIELY, E. R., *Surveying Instruments, their history and use* (New York, 1947)
91. STATIUS, *Silvae* I. 5. 26; PLINY, *Nat. History* 31. 41
92. *De Aqua Ductu* II, 78—87
93. VITRUVIUS, *De Architectura* VIII. 6. 1—2
94. PLINY, *Nat. History* 31. 37
95. *De Aqua Ductu* I. 35
96. *De Aqua Ductu* I. 36
97. FORCHHEIMER, PH., *Wasserströmungen* (In: GEIGER und SCHEIL, *Handbuch der Physik*, Bd. VII, 1927, 192)
98. TANNERY, P., *Frontine et Vitruve* (*Rev. de Philol.*, vol. 21, 1897, 118)
KRETSCHMER, FR., *Rohrberechnungen und Strömungsmessung in der altrömischen Wasserversorgung* (*Z.V D.I.*, vol. 78, 1934, 19—22)
KRETSCHMER, FR., *Römische Wasserhähne* (Jahrb. Schweiz. Ges. Urgesch. vol. XLVIII, 1960/61, pags. 50—62)
99. *De Aqua Ductu*, II. 112—113
100. PLINY, *Letters* IX. XIX (edit. Loeb Classical Library, New York, 1924)
101. KOBERT, R., *Chronische Bleivergiftung im Klassischen Altertum.*
DIERGART, *Beitr. aus der Geschichte der Chemie*, Leipzig, 1909) (In:
102. *De Architectura* VIII. 4
103. TRILLAT, A., *Sur un procédé colorimétrique utilisé par les Romains pour caractériser les eaux douces* (*C.R.*, 1916, 486—488)
104. BOLLEN, A., *De hoedanigheden van het drinkwater volgens de oude geneeskunde* (*Thèse de lic.*, Univ. de Louvain, 1942, 1943)
BAKER, M. N., *The Quest for Pure Water* (New York, 1948, 1—8)
SQUAZZI, FR., *L'arte idro-sanitaria degliantichi.* (Filelfo, Tolinteno, 1954)
105. HERODOTUS, I. 188
106. ARISTOTLE, *Meteorologica*, II. 3; Hist. Anim. IX. 2

107. Forbes, R. J., *Short History of the Art of Distillation* (Leiden, 1948, 14—15)
108. Müller, R. et alia, *Die Umschau*, vol. 38, 1934, 744, 827
109. *De Architectura* VIII. 6. 15
110. II Kings, 2. 21—22 (see also Ratner, *Die Trinkbarmachung ungeniessbaren Wassers in der Bibel, Hyg. Rdschau*, vol. 20, 1910, 190—191)
111. Exodus 15, 25. (See also Driver, G. R., *The "Waters of Bitterness"* (Syria vol. XXXIII, 1956, 70—78)
112. Heron, *Pneumatics* I. chap. 32
113. Mengeringshausen, M., *Die häusliche Verwendung von Wasser und Abwasserwirtschaft im Mittelalter* (*Technikgeschichte*, vol. 25, 1936, 43—56)
Sabine, E. L., *Latrines and cesspools in medieval London* (*Speculum*, vol. 9, 1934 303, 321)
Sabine, E. L., *City cleaning in medieval London* (*Speculum*, vol. 12, 1937, 19—43)
114. Ehlers, G., *Die Wasserversorgung der deutschen Städte im Mittelalter* (*Technikgeschichte*, vol. 25, 1936, 13—24)
115. Thorndike, L., *Sanitation, baths and street cleaning in the Middle Ages the Renaissance* (*Speculum*, vol. 3, 1928, 192—203)
—, *History of Medieval Europe* (New York, 1949, 431—532, 655)
116. Holmes, U. T., *Daily Living in the Twelfth Century* (Madison, 1952, 92, 101 & 285)
117. Parsons, W. P., *Engineers and Engineering in the Renaissance* (Baltimore, 1939, 240—245)
118. Robins, W. F., *The Story of Water Supply* (Oxford, 1946, 115)
Buffet, B. et Evrard, R., *L'eau potable à travers les âges* (Liège, 1950)

INDEX